ENDOCRINOLOGY RESEARCH AND CLINICAL DEVELOPMENTS

ESTROGEN RECEPTORS

STRUCTURE, FUNCTIONS AND CLINICAL ASPECTS

Endocrinology Research and Clinical Developments

Additional books and e-books in this series can be found on Nova's website under the Series tab.

ENDOCRINOLOGY RESEARCH AND CLINICAL DEVELOPMENTS

ESTROGEN RECEPTORS

STRUCTURE, FUNCTIONS AND CLINICAL ASPECTS

GEORGE G CHEN
MICHAEL CF TONG
AND
CA VAN HASSELT
EDITORS

nova
Medicine & Health
New York

Additional color graphics may be available in the e-book version of this book.

Library of Congress Cataloging-in-Publication Data

Names: Chen, George G, editor. | Tong, Michael CF, editor. | Van Hasselt, CA, editor.
Title: Estrogen receptors : structure, functions and clinical aspects / George G. Chen, PHD, Michael C.F. Tong and CA van Hasselt, editors, Department of Otorhinolaryngology, Head and Neck Surgery, Faculty of Medicine, The Chinese University of Hong Kong, Shatin, NT, Hong Kong.
Identifiers: LCCN 2020029133 (print) | LCCN 2020029134 (ebook) | ISBN 9781536182286 (hardcover) | ISBN 9781536182965 (adobe pdf)
Subjects: LCSH: Estrogen. | Estrogen--Receptors.
Classification: LCC QP572.E85 E833 2020 (print) | LCC QP572.E85 (ebook) | DDC 573.4/4--dc23
LC record available at https://lccn.loc.gov/2020029133
LC ebook record available at https://lccn.loc.gov/2020029134

Published by Nova Science Publishers, Inc. † New York

CONTENTS

PREFACE

Estrogen receptors exist in a variety of human organs/tissues of both females and males. They influence almost every aspect of human physiological functions, activities and behaviours. In mammals there are two main estrogen receptors, estrogen receptor alpha (ERα) and estrogen receptor beta (ERβ). Either ERα or ERβ has several isoforms. The function of estrogen receptors is usually activated by their endogenous ligand estrogens, 17β-estradiol mainly. Disturbance in estrogen receptors including their isoforms is known to cause various pathological conditions ranging from reproductive disorders to many types of cancers. However, on the other hand, estrogen receptors in the physiological condition are necessary for the development, growth and maintenance of various organs and tissues and they protect cells from damages.

This 8-chapter book is focused on estrogen receptor structure, functions and clinical significance. Though we are unable to address all aspects of the estrogen receptors, we have analysed and discussed estrogen receptors or their isoforms in some major estrogen receptor-related disorders including cancers, endocrine diseases, respiratory diseases and obesity. The first three chapters address estrogen receptors in breast cancer. Chapter 4 discusses ERα in endocrine disorders including autoimmune thyroid diseases, thyroid cancer, and type 2 diabetes. Chapter 5 analyzes estrogen receptors in lung cancer. Chapter 6 explores estrogen receptors in airway diseases: bronchiectasis, cystic fibrosis, asthma and chronic obstructive pulmonary disease. Chapter 7 summarizes one emerging ERα isoform, ERα36, in various types of cancers. The last Chapter describes estrogen receptors in obesity.

Our understanding of estrogen receptors is still incomplete. More efforts should be devoted to developing mechanistic studies to delineate more precisely the expression, localization, and function of estrogen receptors and their isoforms in different types of cells/tissues. Without these detailed data, the clinical values of estrogen receptors and their isoforms are hard to be determined and estrogen receptor-related target therapies may face a great deal of issues. We hope that the information provided by our book will aid the development toward this direction. We believe, this book will be of great value not only for clinicians interested in estrogen receptor-related diseases but also for basic scientists working in the field of estrogen receptors.

George G Chen, Michael CF Tong and Charles Andrew van Hasselt
Department of Otorhinolaryngology, Head and Neck Surgery
The Chinese University of Hong Kong, Hong Kong

26 May 2020

In: Estrogen Receptors
Editors: George G Chen et al.
ISBN: 978-1-53618-228-6

Chapter 1

THE ROLE OF ESTROGEN RECEPTORS ALPHA IN BREAST CARCINOGENESIS: AN OVERVIEW

Micaela Almeida[1], Mafalda Soares[1], José Fonseca-Moutinho[1,2], Ana Cristina Ramalhinho[1,2,*] and Luiza Breitenfeld[1]

[1]CICS-UBI – Centro de Investigação em Ciências da Saúde, Universidade da Beira Interior, Covilhã, Portugal
[2]Centro Hospitalar Cova da Beira, Covilhã, Portugal

ABSTRACT

Estrogen receptors are divided in two different classes: nuclear estrogen receptors, which can occur in two different isoforms, ERα and ERβ, and membrane estrogen receptors (mERs) (GPER also known as GPR30, ER-X, and Gq-mER), which are mostly G protein-coupled receptors. Nuclear ERα and ERβ mediate the biological effects of estrogens in different tissues. The distribution of these receptors differs from tissues to tissues. ERα is predominantly expressed in uterus, ovarian theca cells, liver and is the most expressed ER isoform in breast epithelium.

* Corresponding Author's Email: cramalhinho@fcsaude.ubi.pt.

ERα protein is codified by ERα gene (ESR1), which is located on chromosome 6q25.1. The eight exons of ESR1 encode the 530 amino acids of ERα, with a molecular size of 54 kDa. ERs are composed by six functional domains: A, B, C, D, E and F. Among these functional domains is H2-terminal domain (NTD), DNA-binding domain (DBD) and ligand-binding domain (LBD). Through LBD, Estradiol (E2) binds to ERs and induces the conversion of the inactive ER to the active form. Estrogens have been clearly identified as carcinogens, the main effects of estrogens are thought to be via stimulating breast-cell proliferation, which increases the chances of proliferation of a potentially mutated cell. ERα activation by estrogens is generally considered responsible for enhanced proliferation in breast cancer in which ERα is expressed in 50 to 80% of cases. Also, several polymorphisms and point mutations in ERα may alter the protein expression and function. Thus, ERα has been considered as an important breast cancer risk factor and as a relevant therapeutical prognostic factor. This chapter provides an overview and discusses the challenges of ERα role in breast carcinogenesis.

INTRODUCTION

Breast cancer is the most frequently diagnosed cancer in women and the most frequent cause of death in women in the majority regions of the world [1]. Breast cancer initiation is generally a consequence of cumulative risk factors, such as single nucleotide polymorphisms (SNPs), acquired mutations, family history, reproductive history and lifestyle habits [2, 3]. The exact etiology of breast cancer is still unknown, however there is a strong evidence that genes of the estrogen pathway may influence breast cancer risk and that estrogens play a critical role in the development, progression of the disease [2, 4]. Endogenous estrogen levels might be critically involved in the progression of breast cancer, once about 70-75% of breast cancers express estrogen receptors (ERs) and respond well to endocrine therapy [2-6]. On this regard, we will further discuss the effect of estrogen receptors, mainly of estrogen receptor alpha (ERα) in human breast cancer.

ESTROGEN RECEPTORS

Estrogens are steroidal hormones that regulate several physiological processes in normal cells including maintenance of reproductive organs, regulation of cardiovascular system, homeostasis of central nervous system, maintenance of bone density and cholesterol levels [7, 8]. Estrogens also play an important role in several pathological processes like cancer, metabolic diseases and osteoporosis [7-9]. The endogenous estrogens are estrone (E1), estradiol (E2) and estriol (E3), being E2 the more predominant estrogen in non-pregnant and pre-menopausal women [6, 10].

The biological effects of E2 are mostly mediated by its binding and activation to estrogen receptors [6, 7, 9, 10]. There are two different classes of estrogen receptors: nuclear estrogen receptors, estrogen receptor alpha (ERα) and estrogen receptor beta (ERβ), and membrane estrogen receptors, the G-protein coupled estrogen receptor (GPER also known as GPR30) [2, 5, 11].

The distribution of steroid receptors varies among tissues being ERα mainly expressed in uterus, ovarian theca cells, breast (being the most expressed ER in breast epithelium), prostate stroma, Leydig cells in testis, epididymis, pituitary, kidney, bone, white adipose tissue and liver [2, 6, 9, 12]. ERβ is mainly expressed in ovarian granulose cells, central nervous system, cardiovascular system, lung, male reproductive organs, prostate epithelium, colon, kidney, bone marrow and immune system [2, 6, 9, 12].

Estrogen Receptors Structure

The estrogen receptors ERα and ERβ are encoded by two distinct genes. The ESR1 gene located at 6q25.1 was cloned for the first time by Walter et al. (1985) and the cDNA was first sequenced by Green et al. (1986) [13, 14]. ESR1 gene contains eight exons that encode ERα, a 66kDa protein, composed by 595 amino-acids [13-15]. The ESR2 gene located at 14q23.2 was first identified by Mosselman et al. 1996 and cDNA was first cloned by

Kuiper et al. 1996, ESR2 contains eight exons that encode ERβ, a 60kDa protein, composed by 530 amino-acids [7, 16, 17].

The ERs are composed by six structural domains (Figure.1) and they share structural characteristics, will focus on describing ERα structure.

The A/B region, encoded by exon 1, shares 17% of homology between ERs, the ligand-independent activation function-1 domain (AF1) part of the NH2-terminal domain (NTD) is involved in the transcriptional activation of target genes and is an essential domain of interactions with co-regulators [7, 9, 12].

The DNA binding-domain (DBD) in the C region is encoded by exons 2, 3 and a part of exon 4, is highly conserved between ERs (97%) and is essential for the specific binding to DNA sequences of target genes, regulating their expression [7, 9, 12].

The D domain, flexible hinge finger, shares 36% of homology, is encoded by exon 4 and links the C domain to E domain, contains the nuclear localization signal (NLS), resulting in the activation of ER signalling in cells, aiding in the translocation of ERα to the nucleus [7, 12, 18].

The exons 4-8 encode for E and F domains, the domain E and F domains shares 56% and 18% of homology, respectively [7]. These domains also called ligand-binding domain (LBD) is a globular region that contains a hormone-binding site, a dimerization interface and a ligand-dependent activation function-2 (AF-2), that has conformational alterations depending on the presence of different ligands [7, 12].

Figure 1. ERα structure. The ERα is a 66kDa protein, with 595 amino-acids, composed by six structural domains. A/B region is involved in the transcriptional activation of target genes, through the ligand-independent activation function-1 domain (AF1) part of the NH2-terminal domain (NTD). The C and D regions have the DNA binding domain and nuclear localization signal, respectively. The E and F regions have the ligand binding domains (LBD) and ligand-dependent activation function-2 (AF-2).

ERα Signalling

Estrogen plays an important role in cell proliferation and differentiation, mainly by binding to its receptors, among them ERα, leading to their activation [19, 20].

In breast cancer, ERα mediates E2 action in the nucleus by two distinct pathways: the genomic and non-genomic pathways (Figure. 2) [7, 12, 21, 22]. The genomic pathway can be divided into estrogen and estrogen response elements (ERE) dependent pathway ("classical" genomic pathway), and estrogen-dependent/ERE-independent pathway ("nonclassical" genomic pathway) and estrogen-independent/ERE-dependent pathway [21].

In the "classical" pathway, E2 binds to ERs, through LBD in the cytoplasm, inducing a conformational change in ERα, enabling ERα dimerization and translocation to the nucleus [12, 22, 23]. The complex ERα-E2, through the DBD of ERα binds to estrogen response elements (ERE), activated ERα triggers the recruitment of co-regulators, through AF-1 and AF-2 domains, that modulates the transcriptional activity [12, 20-26]. Alternatively, activated ERα, by E2, can recruit and bind, through protein-protein interactions, to transcription factors (TF) such as SP1, nuclear factor-κB (NF-κB) and AP1, at their respective response element (RE), stimulating the transcription of growth and survival genes [4, 20-22, 24, 25].

There is also a third pathway, E2-independent/ERE-dependent ERα, a crosstalk between ERα and growth factors induces the activation of kinases, like mitogen-activated protein kinase (MAPK) pathway [4, 21, 23, 25]. Phosphorylated ERα translocates to the nucleus and initiates its transcriptional activity, binding to EREs and recruiting co-regulators [22, 23, 25].

The non-genomic pathway, is not well understood as the genomic pathway, but is largely studied for ERα [12, 25]. In this signalling pathway, E2 acting at the membrane, stimulates the ERα located in the cytoplasm, after that, it initiates a signalling cascade, leading to rapid physiological responses [12, 22, 25, 27]. The ERα-E2 complex activates kinases, such as Src kinase and phosphatidylinositol 3-kinase (PI3K) pathway, activating the signalling cascade of mitogen-activated protein kinase (MAPK) and protein kinase B (AKT), these pathways enable the activation of other transcription factors, facilitators of cell growth and proliferation [20-23, 25, 26, 28].

These genomic and non-genomic mechanisms of ERα signalling are presented separately, however the signalling pathways are strongly related, enforcing the central role of ERα in cell development and differentiation [23, 25].

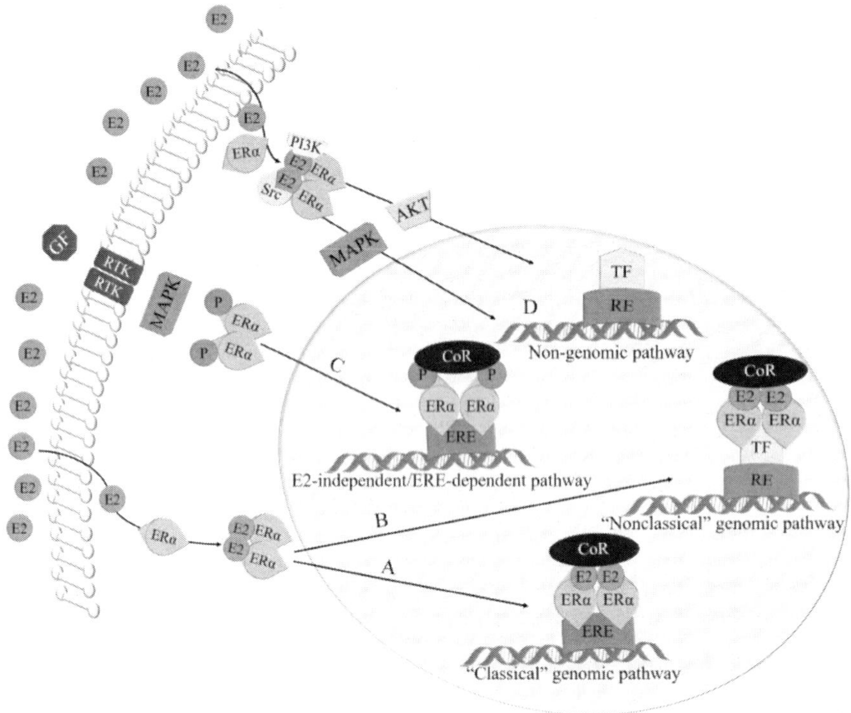

Figure 2. ERα signalling pathways. A) is the "classical" genomic pathway, E2 binds to ERs, the complex binds to ERE and co-regulators are recruited. B) "nonclassical" genomic pathway, E2 binds to ERs, through protein-protein interactions the complex binds to transcription factors (TF) at their respective response element (RE). C) estrogen-independent/ERE-dependent pathway, a crosstalk between ERα and growth factors induces the activation of kinases, phosphorylated ERα binds to EREs and recruiting co-regulators. D) non-genomic pathway, ERα-E2 complex activates kinases, through MAPK and AKT signalling cascades, transcription factors and their regulators enable the activation of other transcription factors.

ESR1 Mutations

The risk for the development of breast cancer is long been associated to lifetime exposure to E2 or estrogenic compounds, mainly through E2/ERα dysregulation in the genomic and non-genomic pathways, that might lead to breast carcinogenesis [2, 4, 21, 23, 25, 26, 29, 30]. The genetic component should not be discarded once mutations in ERα gene (*ESR1*) may alter the protein expression and function.

Table 1. Studies correlating *ESR1* polymorphisms and the risk of breast cancer, worldwide

SNPs	Authors	Year	Population	Relation to Breast Cancer
rs1062577	Chen et al. [33]	2016	Chinese	Increased risk of breast cancer
	Dehghan et al. [44]	2017	Iranian	
rs2046210	Wang et al. [39]	2014	Chinese	Increased risk of breast cancer
	Thanh et al. [46]	2018	Vietnamese	
	Mizoo et al. [47]	2013	Japanese	
	Campa et al. [48]	2011	European	
rs2077647	Son et al. [34]	2015	Korean	Increased risk of breast cancer
rs2234693 (PvuII)	Wang et al. [39]	2014	Chinese	Increased risk of breast cancer
	Carrillo-Moreno et al. [38]	2018	Mexican	Increased risk of breast cancer
	Gonzalez-Mancha et al. [37]	2008	Spanish	Moderately correlated with breast cancer risk
	Ramalhinho et al. [2]	2013	Portuguese	No association with breast cancer risk
rs2228480 (T594T)	Wang et al. [39]	2014	Chinese	Increased risk of breast cancer
	Kallel et al. [42]	2009	Tunisian	
	Curran et al. [43]	2001	Australian	
rs2273206	Son et al. [34]	2015	Korean	Increased risk of breast cancer
rs2881766	Son et al. [34]	2015	Korean	Increased risk of breast cancer
	Chen et al. [33]	2016	Chinese	Increased risk of breast cancer
	Dai et al. [60]	2019	Chinese	Associated with lymph node metastasis and ER expression and decreased breast cancer risk
rs926778	Son et al. [34]	2015	Korean	Increased risk of breast cancer
rs9340799 (XbaI)	Wang et al. [39]	2014	Chinese	Increased risk of breast cancer
	Ladd et al. [40]	2008	Dutch	No correlation with breast cancer risk
	Shen et al. [41]	2006	Chinese	No correlation with breast cancer risk
	Ramalhinho et al. [2]	2013	Portuguese	Reduced risk of breast cancer for xx (homozygous mutant) carriers
rs9479118	Chen et al. [33]	2016	Chinese	Increased risk of breast cancer

Several mutations in *ESR1* have been described, among them, single nucleotide polymorphisms (SNPs). The most studied SNPs in *ESR1* are rs2234693 (PvuII), rs9340799 (XbaI) and rs2228480 (T594T) (Table 1) [2, 31, 32].

Other SNPs studied in *ESR1* are rs1062577, rs2046210, rs2077647, rs2273206, rs2881766, rs926778 and rs9479118 (Table 1) [31, 33, 34].

On what concerns to the SNPs rs2234693 (PvuII), rs9340799 (XbaI)), both polymorphisms are located at intron 1 and are thought to be functional polymorphisms [2, 31, 32, 35]. The possible mechanism of action is a change in *ESR1* transcription activity, contributing to carcinogenesis [2, 31, 32, 35].

The PvuII [c.453-397T > C] polymorphism has been suggested to increase ERα transcription or produce ERα isoforms [36]. Several studies have been performed in order to understand the correlation of PvuII and breast cancer risk, however there are conflicting results. Gonzalez-Mancha et al. (2008) performed a study on which 1148 samples of Spanish women were genotyped and PvuII T allele (altered allele) was moderately correlated with the increased risk of developing breast [37]. Ramalhinho et al. (2013) performed a similar study in 228 Portuguese women from Central Eastern Portugal and PvuII genotypes were not associated with breast cancer variation risk, the same research team found similar results during a research performed in 2019 [2]. Carrillo-Moreno et al. (2019) also studied PvuII polymorphism in Mexican population, correlating positively breast cancer risk with PvuII and similar results were found by Wang et al. (2014) [38, 39].

The polymorphism XbaI [c.454-351A>G], was studied by Ladd et al. (2008) in Dutch post-menopausal women and no correlation with breast cancer risk was found, similar results were found by Shen et al. (2006) in a Chinese case-control study [40, 41]. However, Wang et al. (2014) associated XbaI polymorphism with increased risk of breast cancer. In the opposite, Ramalhinho et al. (2013) associated the homozygous mutant genotype with a reduced risk of breast cancer [2, 39].

As previously referred, others found out that polymorphism in *ESR1* is the polymorphism T594T [c.1782G>A], located at exon 8, although the exact functionality is unknown, probably has a role on co-regulators recruitment [31]. The T594T polymorphism was correlated with breast cancer risk in Tunisian, Chinese and Australian women [39, 42, 43].

Other SNPs have been studied in breast cancer, among them, the rs1062577 located in the 3'-UTR of *ESR1* is thought to promote higher ERα expression, leading to higher proliferation rate in breast cancer cells [44, 45]. This polymorphism was studied in Chinese and Iranian population by Chen et al. (2016) and Dehghan et al. (2017), respectively, being correlated with the increased risk of breast cancer in both populations [33, 44]. The rs2046210, located upstream *ESR1* was studied in the Chinese, Vietnamese, Japanese population and European population, all studies correlated rs2046210 with an increased risk of breast cancer [39, 46-48].

Son et al. 2015 also studied SNPs in *ESR1* and their correlation with breast cancer risk, and the research group found a positive correlation between rs2881766, rs2077647, rs926778, and rs2273206, and breast cancer risk in Korean population [34]. The polymorphism rs2881766 was also studied by Chen et al. (2016) and was found positively correlated with breast cancer risk in Chinese population, this research group also found an association between breast cancer risk and the polymorphism rs9479118 [33].

All these results emphasize the need to better understand the risk potentiated by *ESR1* SNPs, the impact on ERα, and consequently in breast cancer and therapeutic outcome.

Along with SNPs, other *ESR1* mutations have been studied, mainly posttranslational modifications (PTM) that affect ER signalling, creating an imbalance between ER degradation and stability, that in turn might have impact in therapeutic resistance and clinical outcome [23, 25, 49, 50].

The most common point mutations in *ESR1* are Y537N, Y537S, D538G and K303R, these mutations are related with constant ER activity [23, 49, 51, 52].

The Y537S mutation was first described by Weis et al. (1996) and has been shown to confer hormone independent activation [25, 53, 54]. The Y537N mutation, first described by Fuqua lab in 1997, has found to promote a constitutive transactivation function to ER [25, 53, 55].

Merenbakh-Lamin et al. (2013) first described the mutation D538G that occurs in LBD and results in higher levels of ERα, which is independent of the ligand [49, 56]. Toy et al. (2017) found Y537S, Y537N and D538G in patient samples with metastatic and endocrine-refractory breast cancer, however these mutations were not found in the primary samples, these mutations probably promote hormone-independent growth and activity, and are acquired after endocrine treatment [49].

On what concerns to K303K mutation, it was first described by Fuqua lab in 1997 [55]. The mutation leads to polyubiquitination of ERα in response to antiestrogens, and to an altered recruitment of co-regulators, which is associated with lymph node metastasis and enlarged tumor size, and consequently lead to poor prognosis [23, 25, 57].

It seems that all these point mutations on ERα affects its functional role and lead to endocrine resistance [25]. Moreover, these mutations are possibly more frequently acquired after treatment, when they were not identified in the corresponding sample of the primary tumor but were present in metastatic ER^+ tumors after treatment [49, 58]. Besides the biochemical and clinical implications of these point mutations, there is another challenge, Ross et al. (2019) refer that ER immunohistochemistry is not predictive of *ESR1* mutation status, in this regard, not only the *ESR1* mutations should be analysed in clinical practice, but therapeutics should also be more personalized regarding the result of the SNPs and point mutations [59].

THERAPEUTICS IN BREAST CANCER

As previously discussed, the majority of breast cancer expresses estrogen receptors. Breast cancer molecular subtypes can be evaluated through immunohistochemistry, and they can be classified in four major subtypes, Luminal A, Luminal B, HER-2 enriched and Triple negative (Table 2), according to the hormone receptors stage and proliferation rate [26, 29, 51, 61]. The molecular subtype Luminal A is $ER^+/PR^\pm/HER\text{-}2^-$, the Luminal B $ER^+/PR^\pm/HER\text{-}2^\pm$, the HER-2 enriched is $ER^-/PR^-/HER\text{-}2^+$, and in the Triple negative, there is no expression of hormone receptors ($ER^-/PR^-/HER2^-$) [26, 51, 61].

Table 2. Molecular subtypes of breast cancer

Molecular Subtype	Immunohistochemistry phenotype
Luminal A	$ER^+/PR^\pm/HER\text{-}2^-$
Luminal B	$ER^+/PR^\pm/HER\text{-}2^\pm$
HER2 enriched	$ER^-/PR^-/HER\text{-}2^+$
Triple negative	$ER^-/PR^-/HER2^-$

We will focus on the therapeutic approach of Luminal breast cancer which expressed ERs on targeted therapy. There are two "gold standard" approaches of endocrine therapy to $ER\alpha^+$ breast cancers, by blocking the activation of ERα, through selective estrogen receptor modulators (SERMs) or selective estrogen receptor degraders (SERDs), or by decreasing estrogen levels, through aromatase inhibition, using aromatase inhibitors (AIs) [49, 51, 62, 63].

Selective Estrogen Receptor Modulators (SERMs)

The selective estrogen receptor modulators are a category of therapeutic agents, acting as antagonists of ERα in the breast, that are used for the prevention and treatment of diseases such as breast cancer [12, 63]. The mechanism of action of SERMs is through estrogen competition. The active metabolite has an high affinity to the LBD, inhibiting the recruitment of co-regulators by AF-2 of ERα, displacing the estrogen and blocking the recruitment of co-regulators, inhibiting the proliferative effect of estrogen [12, 30, 62-64].

The first approved drug by Food and Drug Administration (FDA), in 1997, was Tamoxifen, which a has a long biological half-life [65]. The IBIS-I trial, performed between 1992 and 2001, enrolled women with 37 to 70 years old, from Unite Kingdom, Australia, New Zealand, Finland, Spain, Switzerland, Belgium, and Ireland [66]. The main findings of the trial IBIS-

I is that Tamoxifen offers a long period of protection after treatment cessation [66]. Due to the side effects of Tamoxifen, like endometrial cancer, other drugs have been developed, among them are Raloxifene [67].

Raloxifene is a second generation SERM, a drug with short half-life, approved for breast cancer treatment by FDA in 2007 [68]. The clinical trial MORE, performed between 1994 and 1998, in twenty five countries, verified that it reduces the high risk of breast cancer in postmenopausal women with osteoporosis [69]. Other SERMs such as Bazedoxifene, Lasofoxifene, Toremifene and Ospemifene are also used in breast cancer treatment, however Tamoxifen is the only approved drug for $ER\alpha^{+}$ breast cancers in premenopausal women [62, 63, 70, 71]. Beside the good results on breast cancer therapeutics, *de novo* or acquired treatment resistance has been observed, and remains a challenge for breast cancer treatment [51, 63, 64, 72].

Selective Estrogen Receptor Degraders (SERDs)

The selective estrogen receptor degraders, like SERMs, act through antagonizing ERα, but also through its degradation, disrupting the dimerization and nuclear localization of ER and reducing the cellular levels of ERα [62, 63]. Fulvestrant was approved by FDA in 2002, being the first authorized SERD for breast cancer treatment, specifically for hormone receptor-positive metastatic breast cancer in postmenopausal women [73]. Fulvestrant administration consists in an intramuscular injection, however there are bioavailability limitations, what has lead to several researches, mostly on the development of orally bioavailable formulations [74]. Among the new molecules in clinical trials, there is AZD9496, which was in a phase I clinical trial, between September 2014 and June 2019 [75, 76]. Another molecule under trial is RAD1901, two phase I clinical trials have been performed, in ER–positive, HER2–negative, postmenopausal advanced breast cancer patients [77-80]. Other two molecules were also under phase I clinical trials, namely GDC-0927 and LSZ102, which occurred in December of 2014-November 2019 and April 2016-October 2019, respectively [81,

82]. Probably due to these clinical trials in the next few years, new therapeutic based on SERDs might be approved.

Aromatase Inhibitors (AIs)

The biosynthesis of estrogen, through androgens conversion, is catalysed in local tissue, like adipose breast tissue, by enzymes such as aromatase [62, 71]. Therefore, supressing estrogens production through aromatase inhibition is a therapeutic approach in order to diminish estrogen action and thus tumor growth and proliferation, on these regard, therapeutics with Aromatase Inhibitors (AIs) arises. The therapeutics with AIs are of main importance in postmenopausal women, once estrogen production by ovary is reduced, and the main source of estrogen is breast adipose tissue. This therapeutic approach has two main advantages, it is not associated with endometrial cancer risk and thromboembolic events, unlike SERMs [67]. Although, long-term therapeutic is related with osteoporosis and insomnia [63, 83].

Aromatase Inhibitors might be classified into two major categories, steroidal aromatase inhibitors (SAIs) and nonsteroidal aromatase inhibitors (NSAIs). The Exemestane is the steroidal AI approved in 1999 by FDA, for adjuvant breast cancer treatment in postmenopausal women with $ER\alpha^{+}$ breast cancer with prior treatment with Tamoxifen during 2 to 3 years [84]. The effects of Exemestane are irreversible, preventing androgens to bind to the catalytic site of aromatase [85]. Goss et al. (2011) performed a randomized placebo-controlled trial in 4560 healthy postmenopausal women with 5-year Gail risk score greater than 1.66% and found that Exemestane reduced the risk of invasive breast cancer, and in a 3-years follow-up Exemestane was not associated with serious toxic effects [86].

On what concerns to NSAIs, they interact with aromatase through noncovalent interactions, competing with endogenous ligands [87, 88]. There are two approved drugs for breast cancer adjuvant treatment, Anastrozole and Letrozole, approved by FDA in 2001 and in 1998, respectively [88, 89].

The clinical trial IBIS-II, an international double-blind, randomised placebo-controlled trial, evaluated Anastrozole in postmenopausal women with increased risk of breast cancer, between 2003 and 2012 [90]. Through IBIS-II trial it was verified that Anastrozole reduces the risk of ERα$^+$ breast cancer [90]. Another trial. ATAC, a double-blinded randomised trial, compared treatment with Tamoxifen and Anastrozole, alone or in association, during five years in postmenopausal ERα$^+$ breast cancer patients and verified that women treated with Anastrozole experienced less side effects and less recurrence when compared to Tamoxifen, concluding that Anastrozole should be considered as the preferred initial adjuvant treatment [91].

Relatively to Letrozole, a phase III, double-blinded trial (BIG 1-98) compared during five years the therapy with Tamoxifen and Letrozole alone, in postmenopausal breast cancer women [92]. It was verified that women undertaking Tamoxifen experienced more thromboembolic events and endometrial pathology, while women treated with Letrozole experienced more fractures and cardiovascular events, but the main conclusion is that women undertaking Letrozole have higher disease-free survival [92].

Thus, NSAIs can be more tolerable than SAIs and clinical data indicate that they are more effective in postmenopausal women with ERα$^+$ breast cancer. Beside these positive advances in breast cancer treatment, resistance to endocrine therapy remains a challenge in ERα$^+$ breast cancer.

Therapeutic Resistance

Estrogens and estrogen receptors play a key role in breast cancer development, differentiation and progression. During the time, several efforts have been made in order to counteract the effects of estrogen and estrogen receptors, however side effects and resistance to endocrine therapy lead to a decreased overall survival, and about 40 to 50% of ERα$^+$ breast cancer cases have lower disease-free survival, being a matter of concern [93]. Resistance to endocrine therapy might be due to long periods of exposure, the estrogen deprivation might lead to activation of growth factor pathways inducing cancer progression [94, 95]. Moreover, acquired mutations during treatment might promote resistance to treatment [49, 94].

Several mechanisms have been found related to therapy resistance, among them, point mutations in *ESR1* gene like Y537S and D538G [56]. Merenbakh-Lamin et al. (2013), identified D538G mutation, that leads to a conformational change in LBD, thus resistance to Tamoxifen and Fulvestrant might result from this conformational change in helix 12, which interferes with the binding to the receptor, leading to acquired endocrine resistance [56]. Also the mutation K303R was related to resistance to Anastrozole, this point mutation results in an increased binding between ERα-mutated and PI3K, leading to activation of Akt survival pathways, thus not only endocrine therapy with AIs become more ineffective, as also, PI3K/Akt pathway activation leads to tumor progression [96]. Another study from Fribbens et al. (2016), verified that patients with *ESR1* mutations has worse progression-free survival with Fulvestrant when compared to wild-type, thus, *ESR1* mutations analysis prior to treatment, may lead to the choice of a more effective treatment in order to reduce therapeutic resistance [97]. Guerrero-Zotano et al. (2018), performed a study in order to evaluate the resistance of $ER\alpha^+$ breast cancers in prolonged exposure to Letrozole, and verified that probably treatment with Letrozole remodeled the transcriptional landscape, once an higher number of mutations were found in patients with prolonged treatment with Letrozole [98].

In 2019, Xie et al. performed a retrospective study in order to evaluate the differences in treatment with Fulvestrant and Exemestane, and verified that a better efficacy was fund for Fulvestrant when compared to Exemestane after treatment failure with NSAIs [99].

Pan et al. (2017) performed a meta-analysis from 88 trails in women with ER^+ breast cancer that were submitted to endocrine therapy and were disease-free during the next 5 years [100]. The study evaluated the outcomes during the 5 to 20 years post treatment, and verified that twenty years after diagnosis, women still have a persistent risk of recurrence and death [100].

These studies emphasise the need to develop new SERMs/SERDs and AIs with better pharmacological properties, which should also take into account *ESR1* mutations.

Moreover and as previously referred, SNPs are often related to breast cancer risk, but the impact of SNPs in the tumor profile and in breast cancer treatment is not stablished. Several polymorphisms might promote altered transcriptional activity of ERα, contributing to carcinogenesis [2, 31, 32, 35, 44, 45]. Once there are genetic variants that are considered risk factors and that may influence malignant transformation [101], further studies should be performed in order to understand the pathway involved in this process and consequently genetic heterogeneity should be taken into account when choosing the treatment, in order to promote a more individualized and successful treatment.

Menopausal Hormone Therapy in Breast Cancer Patients

Women naturally enter in menopause, or due to treatment, experience a premature ovarian insufficiency, thus the main source of estrogen is adipose breast tissue, where estrogen is catalysed by aromatase [62, 71, 102].

Menopause hormone therapy (MHT) consists in the therapeutic replacement with exogenous estrogen and/or progesterone in order to avoid side effects of menopause [103]. However, there is a potential that MHT induces tumors in women with no history of breast cancer, also, women with previous history of breast cancer, mainly ERα$^+$ breast cancer, might have a recurrence [102]. These two major considerations are not straightforward, as previously referred, there are several signalling pathways for ERα, each women has an unique genetic profile, resulting in an individual response to MHT, thus generalizing if breast cancer survivors should enter in MHT, or not, is dependent of individual characteristics. In this regard, women individual estrogens characterization should be take into account in the evaluation of the best MHT.

CONCLUSION

In this chapter, we provide an overview and the challenges of ERα in breast cancer carcinogenesis. The ERα signalling pathway along SNPs and point mutations are critical to the response of breast cancer to treatment, which they are intimately related to ERα stability. The ERα status is considered as the main factor to predict the response to endocrine therapy. However, further studies should be performed in order to better understand how individual heterogeneity affects the incidence, development and response of breast cancer to endocrine therapy. Although many clinical trials are carried out with new molecules, and that breast cancer patients may benefit from a more effective treatment, prevention through genetic profiles evaluation should also be concerned.

Along with prevention, there is also a need to perform further studies in breast cancer survivals, due to menopause and respective MHT. Overall, much has been done in the past decades on what was concerned about ERα positive breast cancers, but the progress in genomic and biological technologies will complement the knowledge that we know about ERα, making it possible to prevent and treat women, taking into account its individual characteristics. This is the main challenge for breast cancer treatment and will be the turning point to decrease relapse and increase overall survival.

REFERENCES

[1] Ferlay, J; et al. *Int J Cancer.*, 2019, 144, 1941-1953.
[2] Ramalhinho, A; et al. *Molecular biology reports.*, 2013, 40, 5093-5103.
[3] Samavat, H; et al. *Cancer Lett.*, 2015, 356, 231-243.
[4] Hayes, EL; et al. *Breast Cancer Res.*, 2015, 17, 40.
[5] Hsu, LH; et al. *Int. J. Mol. Sci.*, 2019, 20, 306.
[6] Ranganathan, P; et al. *Front. Endocrinol.*, 2019, 10, 708.
[7] Yaşar, P; et al. *Reprod Med Biol.*, 2017, 16, 4-20.

[8] Kong, E; et al. *Biochem Soc Trans.*, 2003, 31, 56-59.

[9] Jia, M; et al. *Best Pract Res Clin Endocrinol Metab.*, 2015, 29, 557-568.

[10] Hua, H; et al. *Exp Hematol Oncol.*, 2018, 7, 1-11.

[11] Louie, MC; et al. *Am J Cancer Res.*, 2017, 7, 1617.

[12] Begam, AJ; et al. *Bioorg Chem.*, 2017, 71, 257-274.

[13] Walter, P; et al. *Proc Natl Acad Sci U S A.*, 1985, 82, 7889-7893.

[14] Greene, GL; et al. *Science.*, 1986, 231, 1150-1154.

[15] Ponglikitmongkol, M; et al. *EMBO J.*, 1988, 7, 3385-3388.

[16] Mosselman, S; et al. *FEBS Lett.*, 1996, 392, 49-53.

[17] Kuiper, G; et al. *Proc Natl Acad Sci U S A.*, 1996, 93, 5925-5930.

[18] Gonzalez, TL; et al. *Toxicology.*, 2019, 141, 41-48.

[19] Mohamed, A; et al. *Am J Pathol.*, 2013, 183, 1096-1112.

[20] Lattouf, H; Poulard, C; et al. *J. Cancer Res. Ther.*, 2016, 7, 690.

[21] Liu, WJ; et al. *J Cell Biochem.*, 2019, 121, 2756-2769.

[22] Vrtačnik, P; et al. *Biochem Med* (Zagreb)., 2014, 24, 329-342.

[23] Tecalco-Cruz, AC; et al. *Clin Breast Cancer.*, 2017, 17, 1-10.

[24] Hewitt, SC; et al. *Rev Endocr Metab Disord.*, 2002, 3, 193-200.

[25] Le Romancer, M; et al. *Endocr Rev.*, 2011, 32, 597-622.

[26] Feng, Y; et al. *Genes Dis.*, 2018, 5, 77-106.

[27] Levin, ER. *Mol Endocrinol.*, 2005, 19, 1951-1959.

[28] Miyoshi, Y; et al. *Med Mol Morphol.*, 2010, 43, 193-196.

[29] Huang, B; et al. *Mol Cell Endocrinol.*, 2015, 418, 240-244.

[30] Sharma, D; et al. *Chem Cent J.*, 2018, 12, 107.

[31] Anghel, A; et al. *Neoplasma.*, 2010, 57, 306-315.

[32] Sun, H; et al. *J BUON*, 2015, 20, 296-308.

[33] Chen, L, et al. *Gynecol Endocrinol.*, 2016, 32, 553-556.

[34] Son, BH; et al. *J Cancer Res Clin Oncol.*, 2015, 141, 633-645.

[35] Liu, X; et al. *J Cancer*, 2018, 9, 2963.

[36] Herrington, DM; et al. *Circulation*, 2002, 105, 1879-1882.

[37] Gonzalez-Mancha, R; et al. *Med Sci Monit.*, 2008, 14, CR136-CR143.

[38] Carrillo-Moreno, DI; et al. *J BUON.*, 2019, 24, 1927-1933.

[39] Wang, Y; et al. *Breast Cancer Res.*, 2014, 16, 422.

[40] Ladd, AGZ; et al. *Breast Cancer Res Treat.*, 2008, 107, 415-419.

[41] Shen, Y; et al. *Cancer Epidemiol Biomarkers Prev.*, 2006, 15, 342-347.
[42] Kallel, I; et al. *J Biomed Biotechnol.*, 2009, 2009, 753683
[43] Curran, JE; et al. *Int J Cancer.*, 2001, 95, 271-275.
[44] Dehghan, Z; et al. *Gene.*, 2017, 611, 9-14.
[45] Li, N; et al. *Breast Cancer Res Treat.*, 2010, 121, 177-184.
[46] Thanh, NTN; et al. *Genes Genet Syst.*, 2018, 17-00053.
[47] Mizoo, T; et al. *BMC cancer.*, 2013, 13, 565.
[48] Campa, D; et al. *J Natl Cancer Inst.*, 2011, 103, 1252-1263.
[49] Toy, W; et al. *Cancer Discov.*, 2017, 7, 277-287.
[50] Jeselsohn, R; et al. *Curr Oncol Rep.*, 2017, 19, 35.
[51] Szostakowska, M; et al. *Breast Cancer Res Treat.*, 2019, 173, 489-497.
[52] Lei, JT; et al. *J Cancer Metastasis Treat.*, 2019, 5, 38.
[53] Fanning, SW; et al. *Elife.*, 2016, 5, e12792.
[54] Weis, KE; et al. *Mol Endocrinol.*, 1996, 10, 1388-1398.
[55] Zhang, QX; et al. *Cancer Res.*, 1997, 57, 1244-1249.
[56] Merenbakh-Lamin, K; et al. *Cancer Res.*, 2013, 73, 6856-6864.
[57] Berry, NB; et al. *Mol Endocrinol.*, 2008, 22, 1535-1551.
[58] Jeselsohn, R. *Breast Care (Basel).*, 2017, 12, 309-313.
[59] Ross, DS; Zehir, A; Brogi, E; Konno, F; Krystel-Whittemore, M; Edelweiss, M; et al. *Mod Pathol.*, 2019, 32, 81.
[60] Dai, Z; et al. *Cancer Cell Int.*, 2019, 19, 11.
[61] Tsang, JY; et al. *Adv Anat Pathol.*, 2020, 27, 27-35.
[62] Saha, T; et al. *Eur J Med Chem.*, 2019, 177, 116-143.
[63] Pepermans, RA; et al. *Steroids.*, 2019, 152, 108493.
[64] Fanning, SW; et al. *Endocrinology.*, 2019, 160, 759-769.
[65] Jordan, VC; et al. *J Endocrinol.*, 1976, 68, 453-460.
[66] Cuzick, J; et al. *Lancet Oncol.*, 2015, 16, 67-75.
[67] Gunn, J; et al. Springer., 2018, 129-145.
[68] Waters, EA; et al. *Breast Cancer Res Treat.*, 2012, 134, 875-880.
[69] Cummings, SR; et al. *JAMA.*, 1999, 281, 2189-2197.
[70] Jordan, VC. *Ann Surg Oncol.*, 2019, 26, 1-10.
[71] Ball, S; et al. *Prev Med.*, 2019, 105834.
[72] Gaillard, SL; et al. *Gynecol Oncol.*, 2019, 154, 199-206.

[73] Bross, PF; et al. *Oncologist.*, 2002, 7, 477-480.
[74] Patel, HK; et al. *Pharmacol Ther.*, 2018, 186, 1-24.
[75] Weir, HM; et al. *Cancer Res.*, 2016, 76, 3307-3318.
[76] NCT02248090. AZD9496 First Time in Patients Ascending Dose Study., December 17, 2019], Available from: https:// clinicaltrials.gov/ct2/show/NCT02248090.
[77] NCT02650817. Phase IB Study to Evaluate RAD1901 on the Availability of Estrogen Receptor Binding Sites in Metastatic Breast Cancer., December 17, 2019], Available from: https://clinicaltrials.gov/ct2/show/NCT02650817.
[78] Harb, WA; Garner, F; Clarkin, M; Roberge, KA; Harris, AG; Williams, G; Hattersley, G. A phase I study of RAD1901, an oral selective estrogen receptor degrader, in ER–positive, HER2–negative, postmenopausal advanced breast cancer patients., 2015, American Society of Clinical Oncology.
[79] Bardia, A; Kabos, P; Elledge, R; Wang, D; Shen, J; Garner, F; O'Neill, A; Kaklamani, VG. Evaluation of RAD1901, a novel investigational, selective estrogen receptor degrader (SERD), for the treatment of ER-positive (ER+) advanced breast cancer., 2017, American Society of Clinical Oncology.
[80] NCT02338349. A Phase I, Multicenter, Open-Label, Multi-Part, Dose-escalation Study of RAD1901 in Postmenopausal Women With Advanced Estrogen Receptor Positive and HER2-Negative Breast Cancer., December 17, 2019], Available from: https://clinicaltrials.gov/ct2/show/NCT02338349?term=RAD1901&draw=2&rank=3.
[81] NCT02316509. A Study of GDC-0927 in Postmenopausal Women With Locally Advanced or Metastatic Estrogen Receptor Positive Breast Cancer., December 17, 2019], Available from: https://clinicaltrials.gov/ct2/show/NCT02316509.
[82] NCT02734615. Phase I/Ib Trial of LSZ102 Single Agent or LSZ102 + LEE011 or LSZ102 + BYL719 in ER+ Breast Cancers. December 17, 2019], Available from: https://clinicaltrials.gov/ct2/ show/ NCT02734615.

[83] Kidwell, KM; et al. *Cancer*, 2014, 120, 2403-2411.
[84] Hashemi-Meshkini, A; et al. *Arch Med Sci*., 2013, 9, 472-478.
[85] Jones, SA; et al. *Clin. Breast Cancer*., 2000, 1, 211-216.
[86] Goss, PE; et al. *N Engl J Med*., 2011, 364, 2381-2391.
[87] Adhikari, N; et al. *Eur J Med Chem*., 2017, 137, 365-438.
[88] Bao, T; et al. *Eur Oncol Haematol*., 2011, 7, 106-108.
[89] Cohen, MH; et al. *Clin Cancer Res*., 2002, 8, 665-669.
[90] Cuzick, J; et al. *Lancet*., 2014, 383, 1041-1048.
[91] Howell, A; et al. *Lancet*., 2005, 365, 60-62.
[92] Coates, AS; et al. *J Clin Oncol*., 2007, 25, 486-492.
[93] Ma, CX; et al. *Oncology* (Williston Park)., 2009, 23, 133-142.
[94] Flaum, LE; et al. *Cancer Treat Res*., 2018, 173, 141-154.
[95] Ferrari, P; et al. Elsevier., 2019, 393-422.
[96] Barone, I; Cui, Y; Herynk, MH; Corona-Rodriguez, A; Giordano, C; Selever, J; et al. *Cancer Res*., 2009, 69, 4724-4732.
[97] Fribbens, C; et al. *J Clin Oncol*., 2016, 34, 2961-2968.
[98] Guerrero-Zotano, AL; et al. *Clin Cancer Res*., 2018, 24, 2517-2529.
[99] Xie, Y; et al. *Clin Breast Cancer*., 2019, 19, e452-e458.
[100] Pan, H; et al. *N Engl J Med*., 2017, 377, 1836-1846.
[101] Madeira, KP; et al. *Mol Biol Rep*., 2014, 41, 5459-5466.
[102] Deli, T; et al. *Pathol Oncol Res*., 2019, 1-16.
[103] Temkin, SM; et al. *Cancer*., 2019, 125, 499-514.

In: Estrogen Receptors
Editors: George G Chen et al.
ISBN: 978-1-53618-228-6

Chapter 2

MOLECULAR MECHANISMS THAT MODULATE THE ACTIVITY OF ESTROGEN RECEPTOR ALPHA IN BREAST CANCER

Angeles C. Tecalco–Cruz[1,*], Marina Macías–Silva[2] and Josué O. Ramírez–Jarquín[2]

[1]Programa de Ciencias Genómicas,
Universidad Autónoma de la Ciudad de México (UACM),
Mexico City, Mexico

[2]Instituto de Fisiología Celular,
Universidad Nacional Autónoma de México (UNAM),
Mexico City, Mexico

ABSTRACT

Estrogen receptor alpha (ERα) has a central role in the progression of mammary tumors. More than 70% of all cases of breast cancer are ERα positive (ERα+). ERα is a nuclear receptor that is shuffling between the cytoplasm and the nucleus, acting as a signaling mediator out of the nucleus (in the cytoplasm and the cell membrane), and as a transcriptional regulator

---

* Corresponding Author's Email: angeles.tecalco@uacm.edu.mx.

in the nucleus. Estradiol hormone binds and activates ERα, whereas the ERα phosphorylation induced by growth factors, as epidermal growth factor (EGF) and insulin growth factor (IGF), can also lead to its activation in breast cancer cells. The ERα activity is associated with mammary tumor growth, and consequently, some drugs are directed to control the abundance and/or activity of this receptor. Although these therapeutic strategies function well in some cases, some patients develop resistance to these treatments, which may be associated with the signaling of ERα. In the last years, several studies have decoded different molecular mechanisms that modulate the ERα activity in breast cancer cells via recruiting different transcriptional co-regulators, but also through mechanisms that modulate its stability and the subcellular localization. Interestingly, these mechanisms are also affected by mutations or posttranslational modifications in the ERα. The knowledge of the molecular mechanisms that regulate ERα activity is important in the identification of new biomarkers and therapeutic targets for detecting and controlling the progression of breast cancer.

Keywords: estrogen receptor alpha, breast cancer, gene expression, stability, nuclear export

INTRODUCTION

Breast cancer is one of the biggest health problems in the world. More than 85% of breast cancers in women have no family history of the disease, presenting a multifactorial influence. Only about 5-10% of breast cancers are related to gene mutations inherited from a parent, and the breast cancer genes 1 and 2 (BRCA 1 and 2) are the most commonly mutated. Breast cancer has the highest incidence among human cancer types and represents the first cause of women's death by cancer in the world, according to the International Agency for Research on Cancer (IARC) [1-3].

The 80% of the cases of mammary tumors develop mainly in the mammary ducts being commonly invasive (*invasive ductal carcinoma*). In a minor percentage, the mammary tumors can also develop in the mammary lobules or both mammary structures, lobules and ducts. At the molecular level, there is high heterogeneity in the mammary tumors transcriptomes. Consequently, there is a lack of biomarkers for early detection, prognostic biomarkers, pharmacological targets, and predictors of sensitivity to the treatments used for controlling the progression of breast cancer [4, 5].

By the molecular complexity of mammary tumors, breast cancer is defined as a collection of mammary neoplasms. Importantly, the biomarkers classically used to identify this disease are estrogen receptor alpha (ERα), progesterone receptor (PR), androgen receptor (AR), and human epidermal growth factor receptor 2 (HER2 or ERBB2). More than 70% of all cases of breast cancer express ERα, and they are classified as *ERα+ breast cancer*, and the rest (~30%) that does not express the ERα are classified as *ERα–breast cancer*. ERα is a nuclear receptor that is recognized and activated by estrogenic hormones, mainly by 17 beta-estradiol hormone (E2), functioning as an important transcriptional regulator and a mediator of signaling pathways in breast cancer cells [6, 7].

Additionally, the St. Gallen International Breast Cancer Conference recommends the classification of breast cancers in different subtypes based on the detection of ERα, PR, HER2, and Ki67 (a cellular marker for proliferation) by immunohistochemistry analysis [8]. Hence, there are four subtypes of breast cancer:

I. *Luminal A-like*: ERα+, PR ≥20%, HER2−, Ki67< 20%.
II. *Luminal B-like*: ERα+, PR < 20% and/or HER2+ and/or Ki67≥ 20%.
III. *HER2-overexpression*: ERα−, PR−, HER2+.
IV. *Basal-like*: ERα−, PR−, HER2− (triple negative).

Luminal A is the most common molecular subtype of invasive breast cancers (40-50%). Luminal B subtype tends to have a worse prognosis than luminal A. Moreover, whereas luminal A is likely to benefit from endocrine therapy alone, luminal B tumors may require additional chemotherapy. The HER2-overexpressing subtype (~15% of all invasive breast cancers) is responsive to anti-HER2-targeted therapy, resulting in a better outcome. In contrast, basal-like breast cancer has the poorest prognosis, since this subtype is not a candidate for endocrine therapy [9].

ERα detection in breast tumors is crucial for the selection of endocrine therapies, which include selective estrogen receptor downregulators

(SERD), selective estrogen receptor modulators (SERM), and aromatase inhibitors (AI). SERD control the ERα stability; SERM are regulators of ERα activity, whereas AI stop the production of estradiol hormone in postmenopausal women. Hence, ERα is a direct target for SERD and SERM, which are antiestrogens that bind to the LBD of this receptor; for instance, fulvestrant is a SERD that induces ERα downregulation by the ubiquitin-proteasome system (UPS). In contrast, tamoxifen is a SERM that binds to ERα, blocking the binding of E2 hormone to ERα and favoring the recruitment of transcriptional corepressors, in breast cancer cells [10-12]. However, patients with breast cancer can display a *de novo* resistance or an *acquired* resistance to therapy by mechanisms not completely understood so far [13].

This chapter is focused on the mechanisms controlling ERα actions in breast cancer via the regulation of its stability and subcellular distribution, as well as its implications in the endocrine therapy resistance and progression of breast cancer.

Regulation of ERα Protein in Breast Cancer

ESR1 gene encodes for the nuclear receptor subfamily 3 group A member 1 (NR3A1), a protein of 595 amino acids with a molecular weight of 66 kDa, known as ERα. Structurally, ERα is organized in 4 main domains: two activation function domains (AF-1 and AF-2), one DNA-binding domain (DBD), and one ligand-binding domain (LBD) for E2. The N-terminal region contains the ligand-independent AF-1 domain, followed by the DBD that is important for dimerization and binding to specific DNA sequences [14, 15]. The C-terminal region contains the LBD and the ligand-independent AF-2 domain. After DBD, there is a hinge region that contributes to a functional synergy between AF-1 and AF-2. The AF-1 and AF-2 domains allow ERα interacts with several transcriptional regulators in an E2-independent or E2-dependent manner, respectively (Figure 1A) [16, 17].

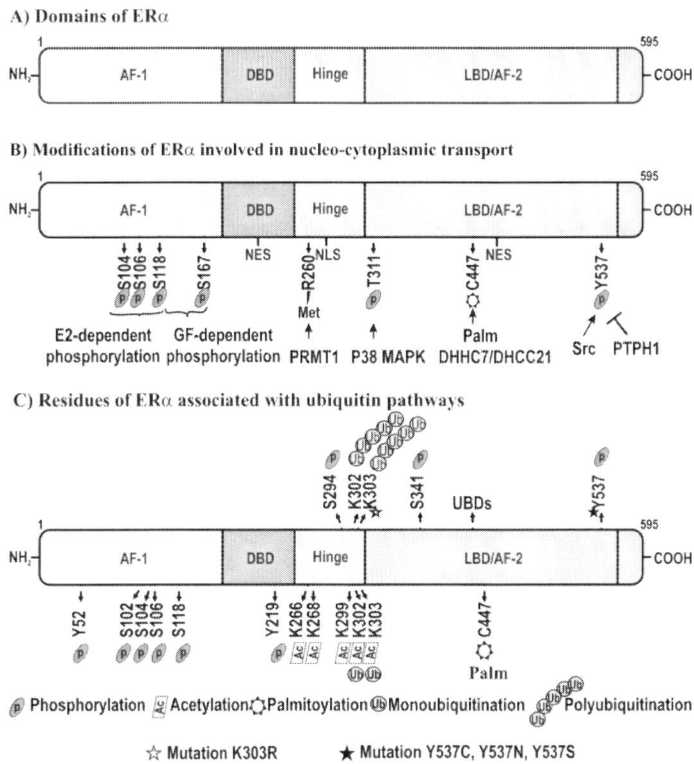

Figure 1. Estrogen receptor α in breast cancer cells. A) ERα is organized in functional domains. The DNA-binding domain (DBD) binds to estrogen response elements (ERE) in enhancers or promoters of target genes. The 17 beta-estradiol hormone (E2) activates the ligand-binding domain (LBD). The hinge domain links LBD and DBD allowing the conformational changes of this receptor. The transactivation domains AF-1 and AF-2 recruit both coactivators and corepressors. B) and C) Some amino acid residues are modified by phosphorylation, acetylation, ubiquitination, and palmitoylation, which are related to ERα regulation via its nucleo-cytoplasmic transport, subcellular localization and stability via ubiquitin pathways. There are interactions between ERα and ubiquitin (Ub) protein: covalent (monoubiquitination and polyubiquitination) and non-covalent via binding to ubiquitin-binding domains (UBD). Some sites of phosphorylation and mutations in ERα that have been identified in breast–cancer biopsy samples are indicated.

ERα also has specific structural motifs that regulate its subcellular distribution, such as the nuclear localization sequences or signals (NLS) and the nuclear export sequences or signals (NES). NLS are located in the hinge region and promote the accumulation of ERα in the nucleus, and the NES are located in the DBD and LBD to mediate the nuclear exclusion of ERα

[18-20]. Additionally, ERα is the target of several posttranslational modifications (PTM) in breast cancer cells, including palmitoylation, phosphorylation, acetylation, methylation, and ubiquitination (Figure 1B and 1C). For example, p300 catalyzes the acetylation of K266 and K268 residues in the hinge region in response to E2, leading to the recruitment of ERα to target gene promoters, whereas the acetylation of K302 and K303 residues decreases the activity of this receptor [21-23].

ERα is methylated at K266 by SET and MYND domain containing 2 (SMYD2) inhibiting its recruitment to the chromatin and its transcriptional activity in the absence of E2. In the presence of E2, ERα methylation at K266 is decreased to promote its acetylation by p300, facilitating its nuclear activity [24]. Also, the methylation at K302 of ERα promotes its binding to target gene promoters to induce transcription [25]. In contrast, ubiquitination at K302 is important for ERα degradation via UPS, and the sumoylation of K266 and K268 promotes the activity of ERα [21, 26].

Interestingly, the AF-1 domain comprises 1-184 aa; from them, 14 are serine residues, and 9 of these serines are sites for phosphorylation. E2 hormone induces the phosphorylation of some of those residues to regulate AF-1 transcriptional activity [27-29]. E2 also induces phosphorylation at the hinge region of ERα at S294 residue, which may facilitate the nuclear import of the receptor in breast cancer cells [30].

Another critical modification for ERα is the palmitoylation, which occurs at C447 residue by the palmitoyl acyltransferases DHHC7 and DHHC21 [31, 32]. The palmitoylated ERα interacts with the caveolin1 protein to be transported and associated with the plasma membrane for rapid extranuclear signaling of E2 [27, 33, 34].

It is important to mention that there are two additional ERα isoforms detected in breast cancer cells named ERα36 and ERα46, with molecular weights of 36 and 46 kDa, respectively. ERα46 lacks the AF1 domain, and ERα36 lacks both AF1 and AF2 domains, containing only DBD and LBD. These isoforms can be localized at the plasma membrane and/or in the cytoplasm. ERα36 is being more studied since it is expressed in both ERα+ and ERα– breast cancer cells and triggers extranuclear signaling pathways in response to E2 to regulate cell proliferation [35-37]. In this review, we focused on describing the molecular mechanisms that modulate mainly the activity of ERα full length (66 kDa) in breast cancer.

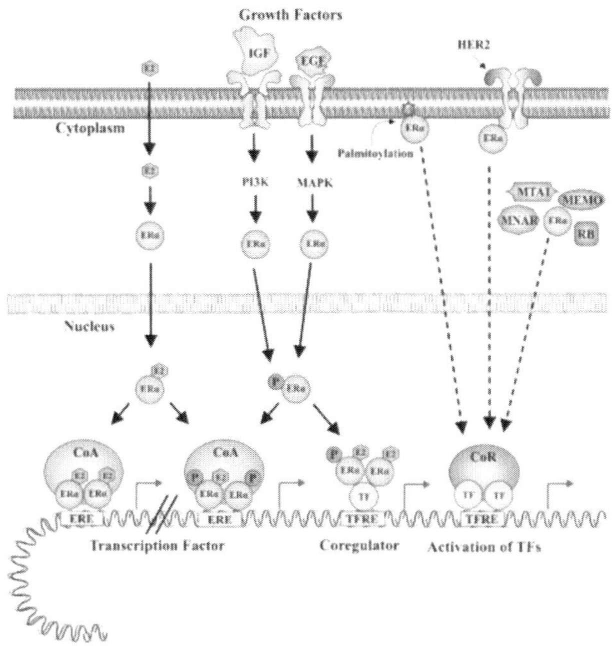

Figure 2. Nuclear and extranuclear signaling of ERα. ERα is recognized by E2 in the cytoplasm and/or nucleus and it forms homodimers that recognize the ERE sequence: AGGTCAnnnTGACCT, in target enhancers and promoters. ERα recruits coregulator (CoR) complexes such as coactivators to induce gene expression. E2 and growth factors (insulin-like growth factor and epidermal growth factor) can induce ERα phosphorylation to promote ERα activity as a transcription factor or coregulator (CoR) for some transcription factors (e.g., AP-1, Sp1, and NF-κB). Cell plasma membrane-associated ERα (via palmitoylation) interacts with transmembranal receptors (e.g., HER2) or with cytoplasmatic proteins (e.g., MEMO, MTA1, and MNAR). Extranuclear interactions of ERα can induce kinase-dependent signaling that could finalize in the activation of some transcription factors.

GENERALITIES ON SIGNALING PATHWAYS OF ERα

In breast cancer cells, ERα is distributed in both the cytoplasm and nucleus. E2 binds ERα in both compartments, triggering different non-genomic and genomic molecular pathways, which seem to be integrated or converge [38]. Here, we will discuss some evidence for nuclear and extranuclear pathways mediated by ERα in breast cancer cells (Figure 2).

Genomic or Nuclear Signaling Pathways of ERα

Nuclear ERα can act as a transcription factor, but also as a transcriptional coregulator of different transcription factors to modulate specific gene expression. As a transcription factor, ERα recognizes and binds directly to a specific palindromic sequence known as the estrogen-responsive element (ERE): 5′-AGGTCAnnnTGACCT-3′. The binding of ERα to ERE determines the selectivity and affinity of the receptor for DNA regulatory regions [39-43]. Briefly, the binding of E2 to the LBD of ERα generates conformational changes to form homodimers that bind to the ERE localized within the enhancers or promoters of E2-responsive genes [39, 44]. In this process, transcription factors named "pioneer factors" are key to open up local chromatin facilitating the binding of ERα to ERE. Then, ERα recruits several coregulators through its AF-1 and AF-2 domains to remodel chromatin structure, either relaxing (coactivators) or compacting (corepressor) the chromatin in order to enhance or reduce the expression of specific target genes, respectively [45].

More than 200 coregulators (coactivators and corepressors) for ERα have been described, modifying the chromatin at specific enhancers and promoters of ERα target genes, and controlling the recruitment of RNA polymerase II complex. These coregulators are also important to coordinate the formation of chromatin loops that affect the genomic organization to allow the regulation of E2-dependent gene transcription [39, 40, 46, 47]. Some coregulators are associated with AF-2, whereas other coregulators are recruited by AF-1 domain; thus, the hinge region acts as a molecular bridge between both domains to synergize their functions. For example, NHERF2 is a coactivator that interacts with AF-1 of ERα and recruits SRC-1 coactivator, and these interactions are increased in response to E2, promoting ERα-dependent gene expression, proliferation, and tumor growth, whereas tristetraprolin (TTP) recruited by AF-1 acts as a corepressor that recruits histone deacetylases (HDAC) to inhibit the expression of ERα target genes, and also inhibits proliferation and tumor progression [48, 49]. Instead, the mediator complex subunit 1 (MED1) and steroid receptor coregulator 3 (SRC3) are critical coregulators that interact with the AF-2 domain of ERα. Furthermore, the regulation of ERα by its coregulators seem to be required for the interactions among enhancer and promoters in the ERα-dependent transcription in breast cancer cells [50-52].

As a coregulator, ERα has no direct interaction with DNA regulatory regions; this receptor indirectly is associated with DNA by its interaction with some transcription factors such as AP-1/c-Jun, ATF-2, NF-κB, p53, SP-1, and STAT [41, 53, 54]. ERα action as a coregulator can modulate gene expression induced by many signaling pathways. ERα also acts as a coregulator that modulates the expression of E2-target late genes [27, 55]. Additionally, ERα transcriptional activity (as a transcription factor or coregulator) can be activated through its phosphorylation induced by E2-dependent, and E2-independent pathways such as those activated by growth factors [56-59]. ERα phosphorylation stimulated by growth factors promotes its transcriptional activity, generating crosstalk between these pathways [60, 61].

Non-Genomic Signaling Pathways of ERα

ERα can be detected at the plasma membrane or cytoplasm of breast cancer cells, triggering the non-genomic or non-nuclear pathways in response to E2 hormone. Hence, palmitoylated ERα localized at the plasma membrane can respond in a fast manner to E2 hormone (response times of 3-15 min) [31, 38, 62, 63]. Also, second messenger production (Ca^{2+}, cAMP, and nitric oxide) can be induced by non-nuclear ERα leading to signaling crosstalk, as well as promoting specific target gene activation. Additionally, ERα interacts with some transmembrane receptors that include GABAB, and mGluR, as well as receptor tyrosine kinases (RTK) of growth factors like EGF and IGF-1 [63, 64].

Noteworthy, GPR30/GPER protein is a G-protein-coupled estrogen receptor unrelated structurally to ERα, localized at the plasma membrane that is also able to recognize the E2 hormone. It has been reported that E2-activated GPR30 generates crosstalk with ERK and AKT-associated signaling pathways in breast cancer cells [65-67]. Moreover, E2-activated GPR30 induces transcription of *FOS* and *CTGF* genes. GPR30 is overexpressed in both ERα+ and ERα– breast cancer cells, favoring cell invasion, and migration [68-70].

ERα Stability in Breast Cancer Cells

ERα Activity is Regulated by Ubiquitin-Dependent Pathways in Breast Cancer Cells

ERα is functionally regulated by mono- and poly-ubiquitination, and some of the ERα-protein interactions occur with ubiquitin, ubiquitin–chains, and mono- or poly-ubiquitinated proteins.

E2 hormone promotes ERα polyubiquitination, inhibiting its monoubiquitination, in breast cancer cells [71]. In this context, ERα monoubiquitination regulates receptor interactions with different partners, its DNA-binding activity, and its stability. Moreover, E2-dependent cellular proliferation seems to be regulated via the monoubiquitination of ERα at K302 and K303 residues [72-75].

The polyubiquitination of ERα is a signal that induces its degradation via the UPS. Intriguingly, RNA polymerase II is not recruited to ERE, when the ERα degradation is inhibited [76]. The use of transcription or proteasome inhibitors abolishes the intranuclear dynamics of ERα [77], and some ERα coactivators are E3-ubiquitin ligases such as E6AP, RNF8, and SKP2 [55, 78-80]. Thus, the ERα-directed transcriptional activity and the cyclical expression of its target genes are interconnected with the ERα protein turnover [76, 81]. Additionally, kinases such as ERK7, CDK11, and PKC catalyze the phosphorylation of ERα to promote its polyubiquitination and degradation via the UPS [82-84].

Ubiquitin-binding domains (UBD) have been described as domains that lack well-defined features but possess regular secondary structures, such as α-helixes and/or Zn^{2+}-fingers [85-87]. The LBD of ERα harbors UBD that mediate the interaction between ERα and ubiquitinated proteins. The key sites of these UBD seem to be the L429 and A430 residues. Interestingly, the mutation of these residues affects E2-dependent ERα transcriptional activity, and also ERα monoubiquitination [88, 89]. Thus, the UBD of ERα emerges as regulators of its interaction with different partners, its monoubiquitination, and its activity, and may have an important role in breast cancer.

ERα-Polyubiquitination and Degradation are Inhibited in Breast Cancer Cells

Some partners of ERα inhibit its polyubiquitination and degradation; these proteins are named as ERα polyubiquitination inhibitor proteins (EPIP) and include MUC1, PIN1, GSK3, LMTK3, c-ABL, RB, RNF31, RNF8, SMURF1, SHARPIN, and TRIM56 [80, 90-99] (Table 1). The downregulation of some EPIP by using RNA interfering strategies increases ERα degradation via the UPS, and decreases its protumoral effects. Thus, in specific EPIP-knockdown cells, the MG132 (proteasome inhibitor) treatment partially restores ERα levels and its effects. Moreover, some EPIPs are mostly expressed in breast cancer tissue compared to normal mammary tissue [80, 97, 98], and some of them may be lost in ERα– breast cancer cells, such as retinoblastoma (RB) [90].

Furthermore, EPIP have specific functions; for example, RNF31, RNF8, SMURF1, SHARPIN, and TRIM56 are E3-ubiquitin ligases; and some of them can monoubiquitinate ERα to stabilize it. LMTK3, c-ABL, and GSK3 are kinases that phosphorylate ERα to promote its stability. Also, there are other mechanisms to promote ERα stability. e.g., PIN1 is an isomerase and an inhibitor of E6AP–ERα interaction; RB is a protein that forms a complex with ERα and HSP90, and MUC-1 functions as a transcriptional coactivator for ERα (Table 1). Thus, multiple proteins seem to interact with ERα to prevent its polyubiquitination and degradation via the UPS, in breast cancer cells.

Table 1. Proteins associated with the stability and subcellular distribution of ERα in breast cancer cells

	Mechanisms	References
Proteins that inhibit ERα polyubiquitination increase its stability in breast cancer		
MUC1	Transcriptional coactivator complex	[95]
PIN1	Isomerization of ERα/Inhibition of ERα ubiquitination by E6AP	[94]
LMTK3	Kinase (via ERα phosphorylation)	[91]
GSK3	Kinase (via ERα phosphorylation)	[92]
cABL	Kinase (via ERα phosphorylation)	[93]
RB	Formation of a complex between RB-HSP90-ERα	[90]
RNF8	E3-Ubiquitin ligase (via ERα monoUbiquitination)	[80]
RNF31	E3-Ubiquitin ligase (via ERα monoUbiquitination)	[97]
SHARPIN	E3-Ubiquitin ligase (via ERα monoUbiquitination)	[98]
SMURF1	E3-Ubiquitin ligase	[96]
TRIM56	E3-Ubiquitin ligase	[99]
DHHC7/ DHHC21	Palmitoylation at C447 residue	[31, 32]
p300	Acetylations probably at K302/K303	[102]
SET7	Methylation at K302 residue	[25]
PEBP4	Interruption of the ERα-Src interaction	[101]
REGγ	ERα stabilization by unknown mechanisms.	[100]
Regulator of the ERα extranuclear distribution and signaling		
Src	ERα phosphorylation at Y537 by Src for its nuclear export	[140]
CRM-1	Exportin for ERα	[20, 139]
MEMO	MEMO associates with ERα and increases its interaction with Src	148]
MNAR	MNAR sequesters ERα in the cytoplasm and the plasma membrane	[149]
MTA1	MTA1 protein sequesters ERα in the cytoplasm	[150]
PTPH1	PTPH1 induces ERα nuclear accumulation by removing its Src-dependent phosphorylation at Y537	[151]
P38 MAPK	ERα is phosphorylated at T311 by p38 MAPK to inhibit its nuclear export.	[152, 153]

On the other hand, phosphatidyl-ethanolamine-binding protein 4 (PEBP4) and REG gamma (REGγ) protein also function as ERα polyubiquitination inhibitors, but they do not interact directly with this receptor. PEBP4 associates with Src to block its interaction with ERα, decreasing ERα phosphorylation, and its subsequent polyubiquitination and degradation [100, 101]. In contrast, REGγ seems to stabilize ERα since REGγ-knockdown cells have low ERα levels, which are increased after MG132 treatment [100].

Additionally, some PTM such as acetylation and palmitoylation, can also stabilize ERα in breast cancer cells. The acetylation of specific lysine residues in the hinge region of ERα by p300 has been associated with an increase in the DNA-binding and transcriptional activities of this receptor [22, 23, 102]. Furthermore, K302 residue is methylated by SET7 to promote ERα stability [25]. ERα palmitoylation at C447 by DHHC7 and DHHC21 induces its localization at the plasma membrane and protects ERα from E2-induced degradation. Mutations in C447 residue block ERα localization at the plasma membrane and facilitate the polyubiquitination and degradation of this receptor in breast cancer cells [31, 32].

The increase of ERα stability appears to enhance the ERα-target gene expression, cell proliferation, and resistance to endocrine therapy of breast cancer cells. Moreover, the accumulation of high ERα protein levels seems to alter the proteolysis of ERα [103-105]. A high ERα concentration activates the transactivation function of ERα through non-classical mechanisms, e.g., E2 signal, growth factors signals or ERα phosphorylation at S104, S106, and S118 residues are not involved in its activation in cells with high levels of ERα. Consequently, there is an atypical ERα recruitment to ERE, p160 coactivator-independent transcription activation, and changes in the magnitude of gene induction, as well as an expansion of new E2 target genes. These effects are called *concentration-inducible ERα function* [106, 107].

ERα Stability and Its Relationship with Endocrine Resistance in Breast Cancer

The Y537 residue in the ERα has a central role in its regulation by ubiquitin-associated mechanisms in breast cancer. The Y537 residue is phosphorylated by Src kinase in order to recruit the E6AP protein, an E3 ligase that is also a coactivator for ERα. E6AP promotes polyubiquitination and degradation of ERα in breast cancer cells, which is inhibited by PIN1 and PEBP4. PIN1 interacts with ERα and inhibit its phosphorylation at Y537 residue, blocking its interaction with E6AP to protect it from degradation. PEBP4 interacts with Src kinase, inhibiting its association with ERα, blocking the phosphorylation and degradation of this receptor. Metastatic mammary tumors from patients with endocrine resistance display mutations in Y537 residue, which may affect the degradation of ERα [79, 94, 101].

Moreover, the upregulation of EPIP such as cABL, PIN1, RNF8, SHARPIN, MUC1, and LMTK3 is related to endocrine resistance and progression of breast cancer. For instance, MUC1 and ERα form a transcriptional complex that inhibits the ERα polyubiquitination but induces the expression of genes for endocrine resistance such as *Rab31* [108]. Another case is LMTK3 that phosphorylates ERα to inhibit its polyubiquitination and degradation; consequently, the reduction of LMTK3 expression is considered as a prognostic marker for better response to tamoxifen [91, 109, 110]. Thereby, the protection of ERα from polyubiquitination and degradation has been associated with endocrine resistance (Figure 3).

Importantly, the loss or reduction of the ERα expression also limits the effect of tamoxifen, which then contributes to endocrine therapy resistance. This mechanism is contrary to the ERα stability-associated endocrine resistance. It has been proposed that some proteins promote the ERα downregulation in breast cancer; for example, high levels of CUEDC2 have been detected in mammary tumors with resistance to tamoxifen. The CUEDC2 protein levels show an inverse correlation with the ERα levels since CUEDC2 induces ERα polyubiquitination and degradation via the proteasome [111, 112]. Moreover, RB-knockdown breast cancer cell lines display an increased polyubiquitination and degradation of ERα, which is inhibited by MG132 treatment, restoring the ERα protein levels in these cells [90]. Hence, resistance to tamoxifen in ERα– mammary tumors has been linked with the deficient expression or dysfunction of RB [113, 114].

Because resistance to tamoxifen is common in breast cancer, it has been considered that fulvestrant is one of the SERD that can be used as a potential therapeutic alternative. Fulvestrant promotes ERα polyubiquitination for its downregulation via UPS. As a result, fulvestrant treatment causes tumor growth reduction. Novel SERD are being studied, such as AZD9496, bazedoxifene, and RAD1901 [115-117], as well as GDC-0810, ZB716, and LSZ102 [118-120]. Despite these findings, so far, fulvestrant is the unique SERD approved for clinical use, as first-line endocrine therapy, and as an alternative for cases with resistance to tamoxifen and aromatase inhibitors [121]. However, the mechanisms that promote or protect ERα from polyubiquitination and degradation may have drawbacks like the development of SERD resistance. A recent study showed that the mediators of the Hippo pathway like large tumor suppressor kinases 1 and 2 (LATS1 and 2), induce indirectly ERα ubiquitination and degradation, and LATS downregulation decreases sensitivity to fulvestrant in ERα+ breast cancer cells. Nevertheless, patients with breast cancer ERα– with short relapse-free survival display high levels of LATS1/2 [122].

Molecularly, the K302 and K303 residues of ERα can be central in the regulation of ERα stability, since these sites may be the target of polyubiquitination induced by fulvestrant and E2 [21]. The K302 and K303 residues are also modified via monoubiquitination by the BRCA-1/BARD1 complex, which is associated with ERα stability at basal conditions [21, 72, 75]. Additionally, SHARPIN-induced monoubiquitination of ERα at those sites may also inhibit its polyubiquitination and degradation [98]. Furthermore, mainly K303 residue is a target of polyubiquitination, monoubiquitination and acetylation in breast cancer cells, suggesting a competition between these modifications for the same site to regulate the ERα stability and activity [21, 25, 72, 75, 102]. These events may be linked to the progression of breast cancer, considering that mutations at K303 residue are displayed in premalignant breast lesions [123, 124].

ERα DISTRIBUTION AND MOLECULAR MECHANISMS IN BREAST CANCER CELLS

ERα has nuclear and extranuclear signaling pathways in breast cancer. Hence, the subcellular distribution and the mechanisms for the nuclear import and export of ERα result critical in the progression of mammary tumors and their response to endocrine treatments.

Notably, karyopherins are proteins that transport molecules from the cytoplasm to the nucleus (nuclear import) and from the nucleus to the cytoplasm (nuclear export) and are called importins and exportins, respectively. For these processes, the importins/cargo or exportins/cargo complexes cross the nuclear pore that is composed of ~30 different nucleoporins and embedded in the nuclear membrane [125, 126]. Importantly, the N-terminal region of importins and exportins contains a Ran-GTP-binding domain. An importin-ERα complex interacts with Ran-GTP in the nucleus, promoting ERα dissociation and release into the nucleus [127, 128]. For nuclear export, the exportin/ERα/Ran-GTP complex requires the hydrolysis of Ran-GTP into Ran-GDP to promote the ERα transport, dissociation, and release into the cytoplasm [129, 130].

Nuclear Import of ERα in Breast Cancer Cells

ERα has a canonic NLS in its hinge domain that is recognized by a heterodimer importin-α/-β [19]. The chaperones HSP90 and HSP70 are elements for the ERα cytoplasmic retention and stability, and the E2 binding induces conformational changes in ERα, exposing its NLS and promoting its dissociation from chaperones for nuclear import, just 10–30 min after E2 stimulation [131-133]. Dynein light chain 1 (DLC1) is a protein that facilitates the nuclear transport of ERα. DLC1 has a chaperone-like activity, and it is upregulated in response to E2 to induce ERα nuclear accumulation, increasing its transcriptional function in breast cancer cells [134].

Growth factors, tamoxifen, and fulvestrant are also associated with ERα subcellular distribution. EGF and IGF1 induce the phosphorylation of ERα, increasing its E2-independent transcriptional activity, as well as, ERα nuclear localization. [56, 57, 59, 135]. ERα is enriched in the nucleus after 60 min of treatment with EGF (via MAPK pathway) or 90 min with IGF1 (via phosphatidylinositol 3-kinase (PI3K) pathway) [135, 136]. Also, tamoxifen induces conformational changes and nuclear accumulation of ERα, promoting a diffuse pattern of ERα detected in the nucleus and associated with repression of its target genes in breast cancer cells [12, 137]. In contrast, fulvestrant treatment stimulates an association between ERα and CK8 and CK18 nuclear matrix-intermediate filament proteins, resulting in nuclear retention of the receptor followed by its degradation via the UPS [138].

Nuclear Export of ERα

Leucine–rich sequences named NES are recognized by exportins to mediate proteins distribution out of the nucleus. ERα lacks classical or canonical NES, but contains sequences similar to NES (non-canonical NES) localized in the DBD and LBD (residues 444–456), which are recognized by CRM-1 exportin to promote ERα nuclear export [18, 20]. Through the construction of ERα tagged with a green fluorescent protein (GFP-ERα) was demonstrated that the nuclear export of ERα is inhibited by LMB (a CRM-1 inhibitor) via the inhibition of ERα/CRM-1 interaction. Moreover, E2 promotes the interaction between ERα and CRM-1 after 1h of treatment [20]. Hence, ERα nuclear export is characterized mainly by being CRM-1-dependent and E2-induced in breast cancer cells.

It has been reported that E2 activates the PI3K/AKT pathway for the transport of the ERα/FKHR complex from the nucleus to the cytoplasm, which is promoted by the AKT-dependent phosphorylation of FKHR [20, 139]. Additionally, since ERα Y537F mutant is accumulated in the nucleus and is unable to interact with CRM-1, it has been proposed that the phosphorylation of Y537 by Src is critical for ERα nuclear export [140].

Furthermore, the E2-induced nuclear export of ERα is reduced by a decrease in the ERα phosphorylation due to the expression of a dominant-negative Src protein or the treatment with SU6656 (a Src inhibitor) [20, 139, 140].

The ERα Nuclear Export is Associated with its Stability in Breast Cancer Cells

ERα subcellular distribution may also depend on its stability, since ERα degradation has been described in the nuclear matrix in response to fulvestrant, in both nucleus and cytoplasm in response to E2 treatment, and in the cytoplasm without any stimulus. Moreover, an interconnection between the subcellular distribution and degradation of the ERα is shown by the Src-dependent phosphorylation at Y537, which promotes its nuclear export, as well as its degradation after polyubiquitination by the E3 ubiquitin ligase E6AP in breast cancer cells [79, 140-142].

The subcellular distribution of ERα is also linked with pathways associated with lysosomal degradation. Lysosomes-associated pathways are implicated in the degradation of extranuclear ERα in response to E2 [143]. Dynamin II is required for autophagy-based ERα degradation. [144], and the ERα has been detected in caveolae, where caveolin 1 and 2 regulate the ERα activity and degradation [145]. Moreover, clathrin-heavy chain (CHC) endocytic protein interacts with ERα in the cytoplasm, promoting ERα degradation, as well as the proliferation of breast cancer cells induced by E2 treatment [146].

HSP90 and HSP70 chaperones promote a cytoplasmic distribution of ERα in the absence of E2, whereas RB protein promotes the formation of a chaperone complex with ERα stabilizing it [90]. Likewise, ERα is modified by diverse PTM to promote its localization and its signaling out of the nucleus. For instance, PRMT1, an arginine methyltransferase, methylates ERα at R260, which helps to maintain ERα in the cytoplasm [147]. Palmitoylation of ERα at C447 by DHHC7 and DHCC21 allows ERα localization at the plasma membrane, which is inhibited by E2 treatment. The palmitoylation of ERα also inhibits its E2-induced degradation [31-33].

The ERα Nuclear Export is Associated with Its Non-Nuclear Signaling Pathways in Breast Cancer Cells

ERα associates with proteins that promote its extranuclear accumulation and its signaling; for example, mediator of ErbB2-driven cell motility (MEMO) associates with ERα and increases the interaction between Src and ERα, enhancing the phosphorylation at Y537 residue in the receptor in order to induce proliferation and migration of breast cancer cells [148]. Modulator of non-genomic actions of the estrogen receptor (MNAR) protein interacts with ERα, sequestrating it in the cytoplasm and at the plasma membrane to promote its extranuclear activities [149]. Truncated MTA1 (metastasis-associated 1) protein also sequesters ERα in the cytoplasm and promotes MAPK-dependent signaling [150]. Nevertheless, there are mechanisms to block ERα extranuclear distribution; for instance, ERα phosphorylation at Y537 by Src is removed by the protein-tyrosine phosphatase H1 (PTPH1) inducing ERα nuclear accumulation [151]. ERα is also phosphorylated at T311 by p38 MAPK to inhibit the nuclear export of the receptor, favoring its accumulation in the nucleus [152]. In contrast, the inhibition of p38 reduces ERα phosphorylation and nuclear distribution, as well as the proliferation of breast cancer cell lines [153]. Human breast tumors display the phosphorylation of ERα at T311, showing the importance of this mechanism in mammary tumors [152].

Remarkably, the nuclear export of ERα limits its nuclear abundance and its transcriptional activity. However, it has been evidenced that LMB treatment inhibits the nuclear export of ERα, but also decreases its transactivation activity [154]. Thus, it seems that the ERα transcriptional activity requires an active nucleo-cytoplasmic shuttling, suggesting a connection between ERα–associated nuclear and extranuclear signaling pathways.

The Subcellular Distribution of ERα Affects the Sensitivity of the Cells to Endocrine Therapy

In general, the extranuclear localization and function of ERα have been associated with endocrine resistance in breast cancer. For instance, the mutations of Y537S, Y537C, and Y537N have been associated with metastasis and resistance to endocrine therapy for mammary tumors in patients [155, 156]. The phosphorylation of Y537 residue seems to be also related to endocrine resistance. PTPH dephosphorylates Y537, accumulating ERα receptor in the nucleus and increasing its sensitivity to tamoxifen and fulvestrant [151].

Also, it has been reported that high doses of fulvestrant increase the association of ERα with the plasma membrane, as well as its phosphorylation by IGF-1R and MAPK in breast cancer cells with endocrine resistance [157]. The translocation of ERα from the nucleus to cytoplasm promotes its extranuclear signaling, which has been evidenced in MCF-7 overexpressing HER2 or ERBB2 that display resistance to tamoxifen [158-160]. The tamoxifen treatment of MCF-7 cells during six months leads to the activation of MAPK, enhancing the sensitivity to EGF and E2, promoting the extranuclear redistribution of ERα, and causing the inability of cells to respond to tamoxifen. Extranuclear ERα interacts with Src and HER2, increasing the survival of tamoxifen-resistant MCF-7 cells [158, 159, 161].

Interestingly, the treatment with tamoxifen and anti-HER2 monoclonal antibody or with the HER2 inhibitor AG825 of tamoxifen-resistant breast cancer cells has positive effects, such as restoring the nuclear localization of ERα and increasing sensitivity to endocrine therapy [158, 159]. The inhibition of Src activity also restores sensitivity to tamoxifen and stimulates the localization of ERα into the nucleus [161]. Likewise, tamoxifen-resistant cells treated with an inhibitor of Src family named Dasatinib plus the addition of tamoxifen reduce the AKT and ERK1/2-dependent extranuclear signaling, recovering the distribution of ERα into the nucleus and the sensitivity of the cells to endocrine therapy [162]. Tamoxifen resistance is also associated with the overexpression of CRM-1; thus, tamoxifen-resistant breast cancer cells treated with the CRM-1 inhibitor denominated Selinexor in combination with tamoxifen restored their sensitivity to tamoxifen [163]. Thus, the nuclear export of ERα is important for the extranuclear interactions of this receptor, to favor the development of endocrine resistance (Figure 3).

Effect of Y537 Mutations in ERα on Breast Cancer

As previously mentioned, the mutations Y537S, Y537C, and Y537N, have been associated with enhanced metastasis and resistance to endocrine therapy of mammary tumors from patients, whereas the mutation of Y537 residue inhibits ERα degradation and blocks its ability to respond to endocrine therapies [79, 142, 155, 156]. Recently, the large scale next-generation sequencing (NGS) applied in mammary tumors from patients has allowed the detection of several *ESR1* mutations such as Y537S, Y537N, Y537C, D538G, and E380Q, which are mainly localized in the LBD of ERα. For this reason, these mutations are associated with conformational changes in the receptor and with its E2-independent activity, as well as, with acquired resistance to endocrine therapy [164, 165]. Remarkably, these mutations have been detected mainly in metastatic breast cancer using digital droplet PCR of circulating tumor DNA [166]. Interestingly, each *ESR1* mutation seems to affect the expression of a specific group of genes [167]. Also, cis-acting targets (cistrome) for each *ESR1* mutation appears to be specific and different, compared to the cis-acting targets for wild type ERα protein [168].

In addition to alteration in DNA binding sites and target gene transcription, ERα mutants display changes in their associations with other proteins. ERα mutations as Y537C and Y537S can be a result of a long term E2 deprivation and may be detected after cells acquiring resistance [169]. On the other hand, resistance to SERM but also to aromatase inhibitors has been related to ERα mutations such as Y537S/N/C, D538G, E380Q, or S463P. Nevertheless, the effects of SERD (fulvestrant and AZD9496) treatments are not affected by mutations in Y537 residue [165, 167, 170], emphasizing the potential use of SERD as a therapeutic strategy in breast cancer with resistance to tamoxifen and aromatase inhibitors.

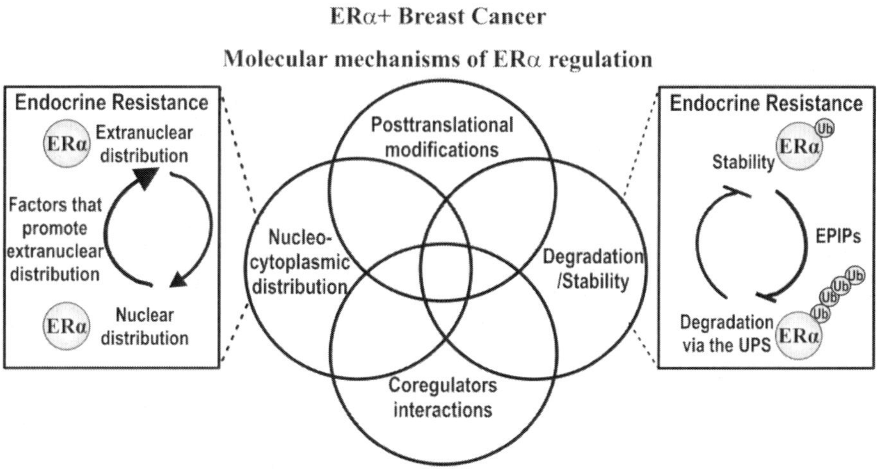

Figure 3. Molecular mechanisms of ERα regulation in ERα+ breast cancer cells. ERα is modulated by its interactions with coregulators and by specific posttranslational modifications that modulate the nucleo-cytoplasmic distribution and stability of this receptor. In an endocrine resistance condition, EPIP can promote ERα stability, and some factors can facilitate the extranuclear distribution of this receptor.

CONCLUSION

ERα is a central factor in breast cancer. Most cases of breast cancer are ERα+, where this receptor has a pro-tumoral activity facilitated by mechanisms that promote its stability, nuclear localization, and transcriptional activity. ERα is a target for SERD and SERM in breast cancer, but the resistance to these treatments occurs. This endocrine resistance seems to be related to ERα enrichment out of the nucleus, promoting its extranuclear signaling, and the sustained inhibition of its degradation. Thus, several molecular connections in and out of the nucleus affect the abundance and activity of ERα in breast cancer, which requires to be explored.

Nevertheless, another way that leads to endocrine resistance is the reduction or loss of ERα. The lack of ERα in breast cancer has been related to epigenetic modifications and miRNAs, and the degradation of ERα protein via the UPS also emerges as one important regulatory mechanism [171]. In ERα– mammary tumors, the absence of this receptor limits the effectiveness of therapy with SERM or SERD. Therefore, the mechanisms for controlling ERα stability have a serious impact on the biology of ERα+ and ERα– breast cancer.

Thus far, fulvestrant (SERD) seems to be a good alternative to face endocrine resistance, but it requires an intramuscular administration and has poor availability. New SERD are required with high affinity and oral availability. Although newer SERD are being investigated, it is important to consider the mechanisms that modulate the ERα stability in this type of cancer to prevent the resistance to these treatments. Many molecular effectors are involved in the modulation of ERα stability, subcellular distribution, and nuclear and extranuclear signaling in breast cancer, suggesting the complex regulation of this receptor, and pointing to ERα regulators as potential targets for treatment.

The knowledge of the molecular mechanisms implicated in the regulation of ERα activity may lead to identify new biomarkers for the endocrine therapy selection and to develop new therapeutic strategies. For instance, there are recent studies on identifying PROTACs (Proteolysis targeting chimeric molecules) that induces ERα proteolysis. Also, selective estrogen receptor covalent antagonists (SERCA) are studied because they can modulate the activity of ERα and its mutants [172, 173] associated with endocrine resistance in metastatic breast cancer [165].

All together indicates that an adequate balance between the abundance and subcellular distribution of ERα may be fundamental in the response of cancer cells to endocrine therapy, and also for controlling the progression of mammary tumors. More integral studies on the modulation of ERα in breast cancer are necessary to help in the detection, control, prognostic, and treatment of breast cancer.

REFERENCES

[1] Bray; F; et al. *CA Cancer J Clin.*, 2018, 68; 394-424.

[2] Ferlay, J; et al. *Int J Cancer.*, 2019, 144; 1941-1953.

[3] Ferlay, J; et al. https://gco.iarc.fr/today; accessed [14 11 2019]. 2018.

[4] Aleskandarany, MA; et al. *Breast Cancer Res Treat.*, 2016, 159, 215-227.

[5] Turashvili, G; et al. *Front Med (Lausanne).*, 2017, 4, 227.

[6] Bertucci, F; et al. *Curr Mol Med.*, 2012, 12, 96-110.

[7] Dai, X; et al. *J Cancer.*, 2016, 7, 1281-1294.

[8] Goldhirsch, A; et al. *Ann Oncol*, 2013, 24, 2206-2223.

[9] Tsang, JYS; et al. *Adv Anat Pathol.*, 2019, 27, 27-35.

[10] Howell, SJ; et al. *Best Pract Res Clin Endocrinol Metab.*, 2004, 18, 47-66.

[11] Wardell, SE; et al. *Biochem Pharmacol.*, 2011, 82, 122-130.

[12] Baumann, CK; et al. *Minerva Ginecol.*, 2009, 61, 517-539.

[13] Osborne, CK; et al. *Annu Rev Med.*, 2011, 62, 233-247.

[14] Kumar, R; et al. *J Amino Acids.*, 2011, 2011, 812540.

[15] Mader, S; et al. *Nucleic Acids Res.*, 1993, 21, 1125-1132.

[16] Metivier, R; et al. *Mol Endocrinol.*, 2001, 15, 1953-1970.

[17] Zwart, W; et al. *J Cell Sci.*, 2010, 123, 1253-1261.

[18] Black, BE; et al. *Curr Biol.*, 2001, 11, 1749-1758.

[19] Burns, KA; et al.; *J Biol Chem.*, 2011, 286, 12640-12649.

[20] Lombardi, M; et al. *J Cell Biol.*, 2008, 182, 327-340.

[21] Berry, NB; et al.; *Mol Endocrinol.*, 2008, 22, 1535-1551.

[22] Kim, MY; et al. *Mol Endocrinol.*, 2006, 20, 1479-1493.

[23] Wang, C; et al. *J Biol Chem.*, 2001, 276, 18375-18383.

[24] Zhang, X; et al. *Proc Natl Acad Sci U.S.A.*, 2013, 110, 17284-17289.

[25] Subramanian, K; et al. *Mol Cell.*, 2008, 30, 336-347.

[26] Sentis, S; et al. *Mol Endocrinol.*, 2005, 19, 2671-2684.

[27] Acconcia, F; et al. *Cancer Lett.*, 2006, 238, 1-14.

[28] Atsriku, C; et al. *Mol Cell Proteomics.*, 2009, 8, 467-480.

[29] Britton, DJ; et al. *J Am Soc Mass Spectrom.*, 2008, 19, 729-740.

[30] Held, JM; et al. *Mol Cancer Res.*, 2012, 10, 1120-1132.

[31] Pedram, A; et al. *Mol Biol Cell.*, 2012, 23, 188-199.

[32] La Rosa, P; et al. *Mol Endocrinol.*, 2012, 26, 762-774.

[33] Acconcia, F; et al. *Mol Biol Cell.*, 2005, 16, 231-237.

[34] Acconcia, F; et al. *Front Physiol.*, 2011, 2, 30.

[35] Li, L; et al. *Proc Natl Acad Sci U.S.A.*, 2003, 100, 4807-4812.

[36] Wang, Z; et al. *Biochem Biophys Res Commun.*, 2005, 336, 1023-1027.
[37] Wang, Z; et al. *Proc Natl Acad Sci U.S.A.*, 2006. 103, 9063-9068.
[38] Acconcia, F; et al. *IUBMB Life.*, 2003, 55, 145-150.
[39] Hah, N; et al. *Genome Res.*, 2013, 23, 1210-1223.
[40] Hervouet, E; et al. *Epigenetics.*, 2013, 8, 237-245.
[41] Marino, M; et al. *Curr Genomics.*, 2006, 7, 497-508.
[42] Metivier, R; et al. *Cell.*, 2003, 115, 751-763.
[43] Kulakosky, PC; et al. *J Mol Endocrinol.*, 2002, 29, 137-152.
[44] Hah, N; et al. *Mol Cell Endocrinol.*, 2014, 382, 652-664.
[45] Manavathi, B; et al. *Front Cell Dev Biol.*, 2014, 2, 34.
[46] Liu, MH; et al. *Mol Cell Endocrinol.*, 2014, 382, 624-632.
[47] Rafique, S; et al. *Genome Biol.*, 2015, 16, 145.
[48] Barrios-Garcia, T; et al. *Mol Genet Metab Rep.*, 2016, 7, 20-26.
[49] Meneses-Morales, I; et al. *Nucleic Acids Res.*, 2014, 42, 6885-6900.
[50] Joseph, C; et al. *Br J Cancer.*, 2018, 118, 1142-1151.
[51] Panigrahi, AK; et al. *Mol Cell.*, 2018, 70, 679-694.
[52] Zhang, X; et al. *Mol Cell.*, 2005, 19, 89-100.
[53] Bjornstrom, L; et al. *Nucl Recept.*, 2004, 2, 3.
[54] Frasor, J; et al. *Cancer Res.*, 2009, 69, 8918-8925.
[55] Zhou, W; et al. *Oncogene.*, 2014, 33, 2341-2353.
[56] Kato, S. Breast *Cancer.*, 2001, 8, 3-9.
[57] Lee, AV; et al. *Clin Cancer Res.*, 2001, 7, 4429s-4435s: discussion 4411s-4412s
[58] Schiff, R; et al. *Clin Cancer Res.*, 2004, 10, 331S-336S.
[59] Smith, CL. *Biol Reprod.*, 1998, 58, 627-632.
[60] Le Romancer, M; et al. *Endocr Rev.*, 2011, 32, 597-622.
[61] Trevino, LS; et al. *Trends Endocrinol Metab.*, 2013, 24, 515-524.
[62] Acconcia, F; et al. *Biochem Biophys Res Commun.*, 2004, 316, 878-883.
[63] Soltysik, K; et al. *J Physiol Pharmacol.*, 2013, 64, 129-142.
[64] Wessler, S; et al. *J Steroid Biochem Mol Biol.*, 2006, 98, 25-35.
[65] Filardo, EJ; et al. *Clin Cancer Res.*, 2006, 12, 6359-6366.
[66] Ignatov, A; et al. *Breast Cancer Res Treat.*, 2011, 128, 457-466.

[67] Poola, I; et al. *Breast Cancer (Auckl).*, 2008, 1, 65-78.
[68] Jiang, QF; et al. *J Steroid Biochem Mol Biol.*, 2013, 138, 314-324.
[69] Maggiolini, M; et al. *J Biol Chem.*, 2004, 279, 27008-27016.
[70] Pandey, DP; et al. *EMBO J.*, 2009, 28, 523-532.
[71] La Rosa, P; et al. *IUBMB Life.*, 2011, 63, 49-53.
[72] Eakin, CM; et al. *Proc Natl Acad Sci U.S.A.*, 2007, 104, 5794-5799.
[73] La Rosa P; et al. *J Steroid Biochem Mol Biol.*, 2011, 127, 223-230.
[74] La Rosa, P; et al. *Cell Signal.*, 2011, 23, 1128-1135.
[75] Ma, Y; et al. *Mol Endocrinol.*, 2010, 24, 76-90.
[76] Reid, G; et al. *Mol Cell.*, 2003, 11, 695-707.
[77] Stenoien, DL; et al. *Nat Cell Biol.*, 2001, 3, 15-23.
[78] Helzer, KT; et al. *J Mol Endocrinol.*, 2015, 54, R151-R167.
[79] Sun, J; et al. *Mol Endocrinol.*, 2012, 26, 1567-1577.
[80] Wang, S; et al. *Biochim Biophys Acta.*, 2017, 1863, 1615-1628.
[81] Metivier, R; et al. *EMBO Rep.*, 2006, 7, 161-167.
[82] Henrich, LM; et al. *Mol Cell Biol.*, 2003, 23, 5979-5988.
[83] Marsaud, V; et al. *Mol Endocrinol.*, 2003, 17, 2013-2027.
[84] Wang, Y; et al. *J Biochem.*, 2009, 145, 331-343.
[85] Dikic, I; et al. *Nat Rev Mol Cell Biol.*, 2009, 10, 659-671.
[86] Husnjak, K; et al. *Annu Rev Biochem.*, 2012, 81, 291-322.
[87] Woelk, T; et al. *Cell Div.*, 2007, 2, 11.
[88] Pesiri, V; et al. *IUBMB Life.*, 2016, 68, 569-577.
[89] Pesiri, V; et al.; *J Cell Sci.*, 2013, 126, 2577-2582
[90] Caligiuri, I; et al. *Oncotarget*, 2013, 4, 875-883.
[91] Giamas, G; et al. *Nat Med.*, 2011, 17, 715-719.
[92] Grisouard, J; et al. *Mol Endocrinol.*, 2007, 21, 2427-2439.
[93] He, X; et al. *Oncogene.*, 2010, 29, 2238-2251.
[94] Rajbhandari, P; et al. *Oncogene.*, 2014, 33, 1438-1447.
[95] Wei, X; et al. *Mol Cell.*, 2006, 21, 295-305.
[96] Yang, H; et al. *J Exp Clin Cancer Res.*, 2018, 37, 24.
[97] Zhu, J; et al. *Oncogene.*, 2014, 33, 4340-4351.
[98] Zhuang, T; et al. *Oncotarget.*, 2017, 8, 77137-77151.
[99] Xue, M; et al. *Oncogenesis.*, 2019, 8, 30.
[100] Chai, F; et al. *Biochem Biophys Res Commun.*, 2015, 456, 534-540.

[101] Liu, H; et al. *J Biol Chem.*, 2010, 285, 21934-21942.
[102] Kim, SH; et al. *Breast Cancer Res.*, 2010, 12, R22.
[103] Alarid, ET; et al. *Mol Endocrinol.*, 1999, 13, 1522-1534.
[104] Preisler-Mashek, MT; et al. *Am J Physiol Endocrinol Metab.*, 2002, 282, E891-898.
[105] Valley, CC; et al. *J Mol Endocrinol.*, 2008, 40, 23-34.
[106] Fowler, AM; et al. *FASEB J.*, 2004, 18, 81-93.
[107] Fowler, AM; et al. *Mol Endocrinol.*, 2006, 20, 291-301.
[108] Jin, C; et al. *PLoS One.*, 2012, 7, e39432.
[109] Stebbing, J; et al. *Breast Cancer Res Treat.*, 2012, 132, 537-544.
[110] Stebbing, J; et al. *Oncogene.*, 2013, 32, 3371-3380.
[111] Musgrove, EA; *Breast Cancer Res.*, 2011, 13, 312.
[112] Pan, X; et al. *Nat Med.*, 2011, 17, 708-714.
[113] Lehn, S; et al. *Cell Cycle.*, 2011, 10, 956-962.
[114] Trere, D; et al. *Ann Oncol.*, 2009, 20, 1818-1823.
[115] De Savi, C; et al. *Cancer Research.*, 2015, 75, 3650.
[116] Garner, F; et al. *Anticancer Drugs.*, 2015, 26, 948-956.
[117] Wardell, SE; et al. *Clin Cancer Res.*, 2013, 19, 2420-2431.
[118] Guo, S; et al. *Oncotarget.*, 2018, 9, 6924-6937.
[119] Lai, A; et al. *J Med Chem.*, 2015, 58, 4888-4904.
[120] Tria, GS; et al. *J Med Chem.*, 2018, 61, 2837-2864.
[121] Deeks, ED. *Drugs.*, 2018, 78, 131-137.
[122] Britschgi, A; et al. *Nature.*, 2017, 541, 541-545.
[123] Fuqua, SA; et al. *Cancer Res.*, 2000, 60, 4026-4029.
[124] Herynk, MH; et al. *Clin Cancer Res.*, 2007, 13, 3235-3243.
[125] Pemberton, LF; et al. *Traffic.*, 2005, 6, 187-198.
[126] Wente, SR; et al. *Cold Spring Harb Perspect Biol.*, 2010, 2, a000562.
[127] Fornerod, M; et al. *Cell.*, 1997, 90, 1051-1060.
[128] Petosa, C; et al. *Mol Cell.*, 2004, 16, 761-775.
[129] Chook, YM; et al. *Curr Opin Struct Biol.*, 2001, 11, 703-715.
[130] Maruvada, P; et al. *J Biol Chem.*, 2003, 278, 12425-12432.
[131] Echeverria, PC; et al. *Biochim Biophys Acta.*, 2010, 1803, 641-649.
[132] Hager, GL; et al. *J Steroid Biochem Mol Biol.*, 2000, 74, 249-254.
[133] Leclercq, G; et al. *Curr Cancer Drug Targets.*, 2006, 6, 39-64.

[134] Rayala, SK; et al. *EMBO Rep.*, 2005, 6, 538-544.
[135] Lu, Q; et al. *FEBS Lett.*, 2002, 516, 1-8.
[136] Takahashi, T; et al. *Endocrinology.*, 2005, 146, 4082-4089.
[137] Shiau, AK; et al. *Cell.*, 1998, 95, 927-937.
[138] Long, X; et al. *J Biol Chem.*, 2006, 281, 9607-9615.
[139] Castoria, G; et al. *Mol Cell Endocrinol.*, 2009, 308, 26-31.
[140] Castoria, G; et al. *Oncogene.*, 2012, 31, 4868-4877.
[141] Arnold, SF; et al. *J Biol Chem.*, 1995, 270, 30205-30212.
[142] Chu, I; et al. *J Clin Invest.*, 2007, 117, 2205-2215.
[143] Totta, P; et al. *PLoS One.*, 2014, 9, e94880.
[144] Totta, P; et al. *Sci Rep.*, 2016, 6, 23727.
[145] Totta, P; et al. *J Cell Physiol.*, 2016, 231, 1219-1225.
[146] Totta, P; et al. *Mol Endocrinol.*, 2015, 29, 739-755.
[147] Le Romancer, M; et al. *Mol Cell.*, 2008, 31, 212-221.
[148] Frei, A; et al. *Oncotarget.*, 2016, 7, 56170-56182.
[149] Vadlamudi, RK; et al. *Hum Pathol.* 2005, 36; 670-675.
[150] Kumar, R; et al. *Nature.*, 2002, 418, 654-657.
[151] Suresh, PS; et al. *Mol Cancer Ther.*, 2014, 13, 230-238.
[152] Lee, H; et al. *Mol Cell Biol.*, 2002, 22, 5835-5845.
[153] Antoon, JW; et al. *Am J Cancer Res.*, 2012, 2, 446-458.
[154] Nonclercq, D; et al. *J Mol Endocrinol.*, 2007, 39, 105-118.
[155] Alluri PG; et al. *Breast Cancer Res.* 2014, 16; 494.
[156] Fanning, SW; et al. *Elife.*, 2016. 5.
[157] Song, RX; et al. *J Steroid Biochem Mol Biol.*, 2010, 118, 219-230.
[158] Pancholi, S; et al. *Endocr Relat Cancer.*, 2008, 15, 985-1002.
[159] Yang, Z; et al. *Clin Cancer Res.*, 2004, 10, 3621-3628.
[160] Shou, J; et al. *J Natl Cancer Inst.*, 2004, 96, 926-935.
[161] Fan, P; et al. *Cancer Res.*, 2007, 67, 1352-1360.
[162] Guest, SK; et al. *PLoS One.*, 2016, 11, e0157397.
[163] Wrobel, K; et al. *Mol Endocrinol.*, 2016, 30, 1029-1045.
[164] Reinert, T; et al. *Curr Treat Options Oncol.*, 2018, 19, 24.
[165] Toy, W; et al. *Cancer Discov.*, 2017, 7, 277-287.
[166] Angus, L; et al. *Cancer Treat Rev.*, 2017, 52, 33-40.
[167] Bahreini, A; et al. *Breast Cancer Res.*, 2017, 19, 60.

[168] Jeselsohn, R; et al. *Cancer Cell.*, 2018, 33, 173-186 e175.
[169] Martin, LA; et al. *Nat Commun.*, 2017, 8, 1865.
[170] Harrod, A; et al. *Oncogene.*, 2017, 36, 2286-2296.
[171] Gajulapalli, VNR; et al. *Biosci Rep.*, 2016, 36
[172] Jiang, Y; et al. *ACS Chem Biol.*, 2018, 13, 628-635
[173] Puyang, X; et al. *Cancer Discov.*, 2018, 8, 1176-1193

In: Estrogen Receptors
Editors: George G Chen et al.
ISBN: 978-1-53618-228-6

Chapter 3

CLINICAL RELEVANCE OF BIOMARKERS IN OESTROGEN RECEPTOR POSITIVE BREAST CANCER

R. M. Parks [*] ***and K. L. Cheung*** [†]***, DM***
School of Medicine, University of Nottingham, Nottingham, UK

ABSTRACT

Some biomarkers have been examined extensively in the context of breast cancer, including the oestrogen receptor (ER) and human epidermal growth factor receptor 2 (HER2). They are now routinely measured and used clinically for selection of primary, (neo)adjuvant endocrine therapy and/or chemotherapy. In this chapter, we will review the literature, including the work of our group, on the analysis of biomarkers (including those that are not routinely measured clinically but may be relevant in ER positive breast cancer e.g., B-cell lymphoma (BCL) 2-protein, Ki-67, and p53). We will specifically look at the relationship of these biomarkers with age and their potential clinical relevance, notably their prognostic value and ability to predict therapeutic response. Furthermore, we will discuss the potential technology of high-throughput measurement of these

[*] Corresponding Author's Email: Ruth.Parks@nottingham.ac.uk.
[†] Corresponding Author's Email: kl.cheung@nottingham.ac.uk

biomarkers at diagnosis for use as tools for clinical assessment and research.

Keywords: oestrogen receptor, epidermal growth factor receptor, biomarkers, immunohistochemistry

ABBREVIATIONS

AIB	Amplified in breast cancer
BCL	B cell lymphoma
BRCA	Breast cancer gene
CK	Cytokeratin
EGFR	Epidermal growth factor receptor
HER	Human epidermal growth factor receptor
LKB1	liver kinase B1
mTOR	Mammalian target of rapamycin
MUC	mucin
NACT	neoadjuvant chemotherapy
ER	Oestrogen receptor
pCR	pathological complete response
PARP	Poly (ADP-ribose) polymerases
PgR	Progesterone receptor
PTEN	Phosphatase and tensin homolog
Tp53	Tumour protein 53
VEGF	Vascular endothelial growth factor

INTRODUCTION

The majority of patients will have core needle biopsy (CNB) performed at diagnosis of breast cancer. This sample is usually analysed on site by a resident histopathologist. It is routine for two types of histological analysis to be completed. Firstly haematoxylin and eosin (H&E) staining is performed to provide information on histological type and grade and secondly, immunohistochemistry to measure the expression of oestrogen receptor (ER), progesterone receptor (PgR) and human epidermal growth factor receptor (HER)-2. These features are important for treatment planning at all stages of therapy (neoadjuvant, primary, adjuvant and advanced) and to help predict recurrence risk and survival outcomes.

Although, ER, PgR and HER2 are now routinely measured, numerous other 'biomarkers' may have a potential role in breast cancer. A 'biomarker' is a characteristic that can be objectively measured and evaluated as an indicator of a normal biological or pathogenic process or pharmacological response to a therapeutic intervention [1]. A number of biomarkers, for example, B-cell lymphoma (BCL)-2, Ki-67 and p53 are well recognised in the literature to be relevant in breast cancer, however they are usually only measured in a research setting.

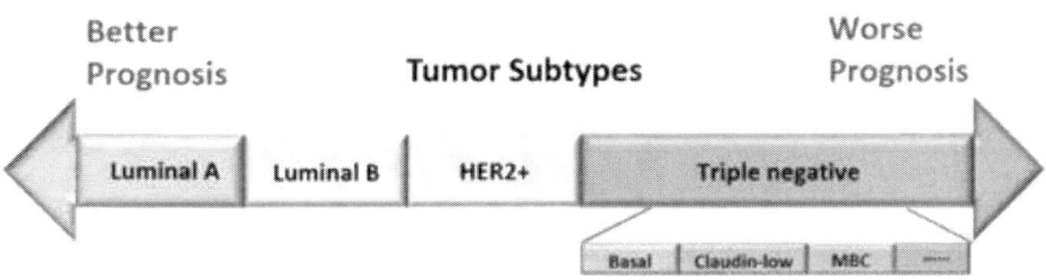

Figure 1. Our current understanding of subtypes of breast cancer [7] (permission to use not required).

We now understand breast cancer to be a heterogeneous complex of diseases, with a spectrum of many subtypes with distinct biological features [2-4] which leads to differences in response patterns to various treatment modalities. Therefore, traditional treatment plans based on routinely measured biomarkers and our current understanding of disease subtypes (Figure 1) may no longer be adequate [5, 6].

In recent years, clinical risk stratification tools such as Adjuvant Online [8] and PREDICT [9] and genomics tool such as Oncotype DX [10] have taken advantage of this concept to assist in decision making regarding adjuvant systemic therapy including chemotherapy. The use of molecular profiling more widely in these settings or in other treatment dilemmas is currently only used in the research setting. A key component of molecular profiling is understanding which biomarkers and in what combination are likely to be most relevant to answer specific treatment questions. In the next section, we will explore individual biomarkers in more detail.

Over two-thirds of all breast cancers are ER positive [11, 12]. Endocrine therapies targeting the ER are utilised at every stage of the treatment pathway including chemoprevention in high-risk patients, primary treatment for patients who are unfit for or decline surgery, in the (neo)adjuvant and advanced settings. However, resistance to endocrine therapy is well recognised but not completely understood. Therefore, it is hypothesised that other biomarkers must pay a role in translating measured ER-positivity to clinically relevant hormone responsiveness. It is for these reasons that we have focused on ER-positive cancer in this chapter. The studies presented in the remainder of this chapter have used immunohistochemistry (IHC) methods to measure biomarkers on patient tumour tissue, unless otherwise stated. These studies include a wide range from the literature, also including those from our group (Nottingham, UK) [6, 13-19]. We possess a large cohort of 1,758 consecutive patients aged ≥70 years with primary operable breast cancer, treated at a single centre with long-term follow-up data available [20]. We have full clinico-pathological data, histological information based on CNB and surgical excision (SE) (if surgery was primary treatment) as well as tissue microarrays (TMAs) constructed from SE. In the same unit, we have a comparative cohort of younger patients (<70 years) again with long-term follow up data, which has been extensively described [21-23].

Biomarkers – Association with Age, Predictive and Prognostic Values, Future Directions

Following on from the well-documented correlation between ER-positivity and age, we will discuss the relationship with age of other biomarkers. Measurement of a biomarker can give both predictive information (how the cancer will respond to a certain therapy), and prognostic information (what the biomarker can tell us about clinical outcome). In order to advance the development of this field, we will discuss future work in the research and clinical settings.

A summary of all biomarkers discussed in given in Table 1.

Table 1. Summary of biomarkers associations with age, efficacy of endocrine therapy and survival

Receptor	Association with age	Association with efficacy of endocrine therapy	Association with survival
ER	Increases with age	Increases	Positivity confers better outcome
PgR	No association confirmed	Increases	Negativity linked to poor outcome
HER2	Decreases with age	Established use of trastuzumab	Positivity confers poor outcome
HER3	Possible increase with age	Linked to resistance	Conflicting evidence, possible poor outcomes
HER4	Possible increase with age	Expression may be linked to resistance	Expression associated with good prognosis
EGFR	Positivity more likely in younger age	Associated with therapy resistance	Poor clinical outcome in terms of survival
VEGF	Conflicting evidence	Anti-VEGF therapy established	Positivity associated with lower survival
Tp53	Chance of mutation increases with age	Linked to therapy resistance	High expression associated with lower survival
BRCA	Possible loss of expression with age	BRCA 1 associated with resistance	Conflicting results
PTEN	Unknown	Downregulation associated with poor response to tamoxifen	PTEN loss predicts worse survival
LKB1	Possible increase with age	Not enough evidence	Conflicting evidence
Ki-67	Lower expression in older women	High expression is an independent predictor of pCR	Positivity associated with worse survival
BCL-2	Conflicting evidence	Better response to tamoxifen	High expression associated with better survival
Cytokeratins	Higher expression of CK5/6, CK5, CK18 in the older population and lower expression of CK 7/8	Unknown – little evidence at present	Basal cytokeratins linked to poor prognosis
MUC1	Unknown	Associated with resistance	High levels associated with poor outcome
AIB1	Unknown	Associated with good response	High levels associated with poor outcome

Hormonal Receptors

If present and activated in breast cancer cells, hormonal receptors will stimulate growth of that cell.

Oestrogen Receptor (ER)

The ERs are a group of proteins found inside cells, which are activated by the hormone oestrogen. Around 70% of breast cancers overall are ER-positive [24].

Association with Age

ER positivity increases with age with the highest proportion of ER-positive cases in patients >65 years of age [25, 26] (Table 2).

Table 2. Percentage of ER-positive by age group in a cohort of 18,586 breast cancer patients by Kerklikowske et al. [25]

Age group (years)	ER-positive cancer (%)
35-39	73.4
40-44	78.0
45-49	76.3
50-54	80.9
55-59	81.6
60-64	83.6
65-69	84.4
70-74	84.3
Overall	*80.5*

Predictive Value

Endocrine agents targeting the ER are effective treatment for breast cancer.

Tamoxifen and raloxifene are approved by the Food and Drug Administration (FDA) for chemoprevention in high risk women [27]. In the International Breast Cancer Prevention Trial [28], 13,388 women deemed to be at an increased risk of breast cancer received tamoxif en or placebo.

Tamoxifen reduced the risk of invasive breast cancer by 49% overall, however there was an increase in side effects.

Neoadjuvant endocrine therapy (NAET) has been used in postmenopausal women with ER-positive breast cancer when chemotherapy is not suitable, either due to tolerability or patient choice, with the aim to improve operability whilst simultaneously treating presumed micrometastases. However, it is noted that the pathological complete response (pCR) of NAET is less than with chemotherapy [29], longer durations of treatment may be required [30] and there is greater disease progression when compared to chemotherapy [31].

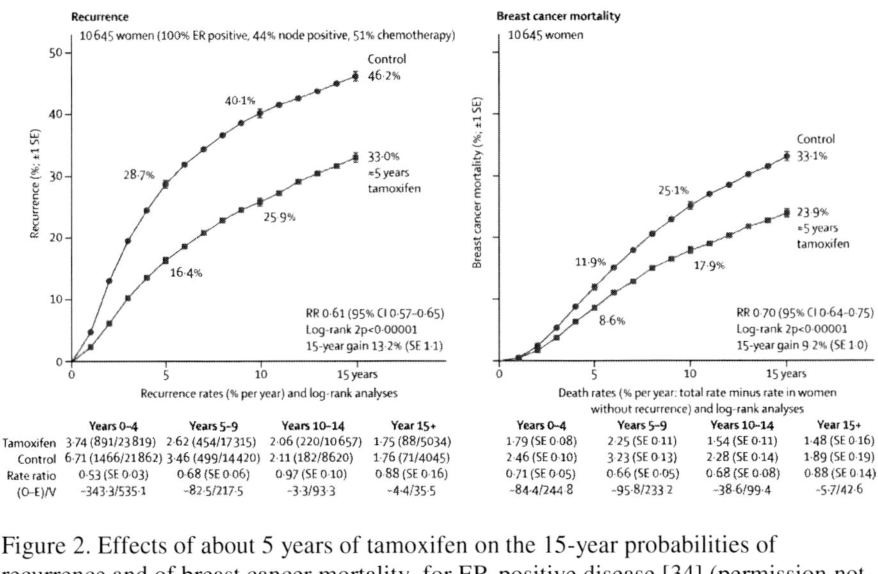

Figure 2. Effects of about 5 years of tamoxifen on the 15-year probabilities of recurrence and of breast cancer mortality, for ER-positive disease [34] (permission not required to use).

Primary endocrine therapy (PET) has been shown to be equivalent in terms of survival outcome compared to surgery [32] and is frequently used in the older population with multiple comorbidity or who decline surgery [33].

Adjuvant endocrine therapy is given for at least 5 years after surgery. The Early Breast Cancer Trialists' Collaborative Group (EBCTCG) meta-analysis published in 2011 [34] showed a 30% reduction in breast cancer mortality (absolute reduction in mortality of 9% maintained for 15 years, following 5 years of adjuvant tamoxifen (Figure 2).

Prognostic Value

Above the age of 40 years, positive ER status is associated with improved breast cancer specific survival (BCSS) and overall survival (OS) [35]. Patients with ER-positive tumours generally have lower risk of mortality compared to those with ER-negative tumours [36], although this is not uniform across all groups. In BRCA1/2 carriers, there is work to suggest that ER-positivity confers a negative outcome [37].

Future Directions

As endocrine therapies are well established as outlined above, current research is involved in combination work with newer targeted agents, for example, cyclin-dependent kinase (CDK) 4/6 inhibitors. This combination is currently utilised in the setting of advanced disease and in clinical trials as adjuvant therapy [38, 39].

Progesterone Receptor (PgR)

PgR is a nuclear receptor, activated by the hormone progesterone. Between 65-75% of breast cancers are PgR positive [40]. Most PgR positive tumours are also ER-positive, so its significance is often downplayed. Progesterone receptor serves as an indicator of a functionally intact nuclear ER pathway because adequate levels of oestrogen and nuclear ER are required to transcribe PgR; therefore it may help predict which patients will respond to hormonal therapy [41].

Association with Age

There is no confirmed relationship between expression of PgR and age to date [6, 26, 42]. Biological characterisation of 14,007 primary breast cancer cases in a single series [43] has suggested that PgR expression may increase with age, but this was not statistically significant.

Predictive Value

In the adjuvant setting, PgR expression is usually associated with greater benefit from endocrine therapy [43, 44]. One study which found conflicting results was the Arimidex, Tamoxifen Alone, or in Combination (ATAC) trial [45]. A total of 9,366 postmenopausal patients with primary breast cancer were found to have significantly improved disease-free survival (DFS) and time to recurrence (TTR) when treated with adjuvant anastrozole compared to tamoxifen. The TTR was longer for patients who were ER-positive and PgR-positive or negative, however the overall survival benefit was greater in the PgR-negative subgroup). The significance of this retrospective subgroup analysis remains controversial.

Prognostic Value

A lack of PgR positivity has been linked to poor survival regardless of age and independent of ER status [46] and increases the risk of relapse and death from breast cancer [47].

In a recent study by Van Asten et al. [48] in 9,647 patients with primary operable ER-positive, HER2-negative breast cancer, absent PgR expression predicted worse outcome in terms of disease free recurrence interval (DFRI) and the group suggested that this effect was more important in high compared with low proliferative tumours.

In a study performed by Albanghali et al. [15] of 536 older women with ER-positive primary breast cancer, PgR expression was associated with better BCSS.

Future Directions

Both the ATAC trial [45] and the study by Van Asten et al. [48] have shown results which require validating in larger prospective cohorts.

Growth Factor Receptors

Breast cancer cells require activated growth factor receptors to proliferate, invade and metastasise. Growth factor receptors transduce extracellular signals through activation of intracellular messengers. The human epidermal growth factor (HER) family is one of the most extensively studied and consists of four members: EGFR, HER2, HER3 and HER4. The receptors are structurally related. The receptors require dimerisation (Figure 3) activation of RAS-MAPK and PI3K-AKT-mTOR downstream signalling pathways [49, 50] (Figure 4).

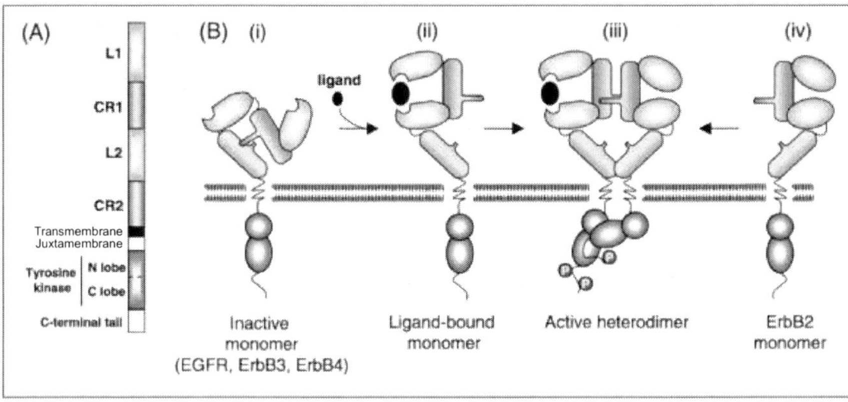

Figure 3. Structure of ErbB receptors [49] (permission to reuse not required).

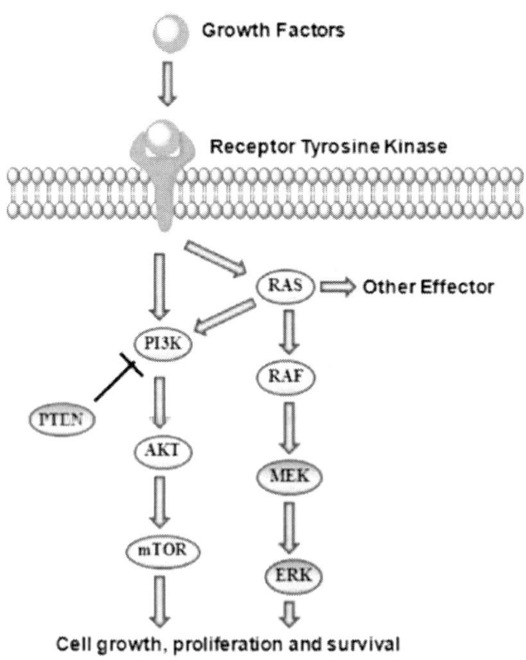

Figure 4. PI3K/Akt/mTOR and Ras/Raf/MEK/ERK signalling pathways [51] (permission to use requested 29/10/2019).

Epidermal Growth Factor Receptor (EGFR)

In an American database of 2,567 breast cancer patients on whom EGFR expression was measured, 18% were EGFR positive [52].

Association with Age

In the same study, EGFR expression was more common in breast tumours in younger (<50 years) compared to older women [52].

Predictive Value

EGFR expression is associated with lower hormone receptor (HR) levels, higher tumour proliferation and HER2 overexpression [52]; which correlated with higher risk of relapse in patients receiving adjuvant treatment. This has been seen in other studies where high levels of EGFR contributed to acquired endocrine resistance with tamoxifen [53].

Prognostic Value

High expression of EGFR correlates with poor clinical outcome in breast cancer [54-56]. In a study by Tsutsui et al. [55] expression of EGFR was analysed in 241 patients with recurrent breast cancer. Patients with positive EGFR expression had worse post-relapse survival than those with negative expression.

Future Directions

There are two types of EGFR inhibitors on the market now; these include tyrosine kinase inhibitors for lung and pancreatitis cancer and monoclonal antibodies for colorectal and head and neck squamous cell carcinoma. For breast cancer however, clinical trials of EGFR inhibitors as monotherapy and in combination with chemotherapy, have not been as successful in universally producing good response rates. Only a handful of trials have noted a good response to EGFR inhibitors in a small number of patients [57, 58]. A phase II trial performed by Johnston et al[58] tested lapatinib, an oral inhibitor of EGFR and HER2 in 45 patients. A clinical response was seen in 50% of HER2-positive patients. Going forward, it may be necessary to stratify patients to enhance the efficacy of EGFR inhibitors and consider combination therapies [59].

Human Epidermal Growth Factor Receptor 2 (HER2)

Approximately 15 of breast cancers have positive expression of HER2 [60, 61]. There is an inverse relationship between HER2 and ER positivity, leading to approximately 10% of ER-positive tumours being HER2-positive and around 50% of HER2-positive tumours being ER-positive [62].

Association with Age

There is lower expression of HER2 with increasing age [6, 60, 63]; HER2 positive tumours are less common over the age of 70 years [63].

Predictive Value

Clinical trials have shown that trastuzumab (Herceptin), a monoclonal antibody targeting the HER2 receptor, significantly improves OS in women with HER2-positive breast cancer and this has been confirmed with long-term follow-up data from the HERA (HERceptin Adjuvant) trial [64]. National Institute for Health and Care Excellence (NICE) guidelines recommend 1 year of adjuvant trastuzumab for stage T1c and above HER2-positive invasive breast cancer in combination with chemotherapy [65].

Retrospective studies have suggested that HER2 overexpression may have a predictive role for response to chemotherapy and endocrine therapy, but results are conflicting [66].

Other HER2 directed therapies approved for clinical use include the monoclonal antibody, pertuzumab, the small molecule kinase inhibitor lapatinib [67] and the toxin-carrying antibody, trastuzumab emtansine [68].

Prognostic Value

HER2 overexpression in general confers an aggressive phenotype and poor patient outcome [69]. Overexpression is associated with increased tumour aggressiveness, increased rates of recurrence and increased mortality in patients with node-positive disease [66].

Future Directions

Pharmacological resistance to trastuzumab and other HER2 directed therapies is recognised. It is hypothesised that HER2-positive breast cancer can be targeted by immunotherapeutic interventions and compounds are currently in the clinical trial setting [50]. One example is immune checkpoint inhibitors, which include CTLA-4 inhibitors, anti-PD-1 and anti PD-L1 antibodies and are currently in Phase I and II trials for early and advanced HER2-positive breast cancer. HER-2 specific peptide-based vaccines have been developed to treat cancers by enhancing the antitumour immune response. One example is Nelipepimut-S (E75) which is currently under investigation in Phase II and III clinical trials [50].

Human Epidermal Growth Factor Receptor 3 (HER3)

Co-expression of HER2 and HER3 is common in breast cancer [70]. The rate of overexpression of HER3 in breast cancer is widely varied according to the literature and ranges from 30-3-75.1% [71].

Association with Age

High expression has been noted in older women (≥70 years) with primary breast cancer, however, no direct comparison to younger patients has been made [6].

Predictive Value

HER3 has been linked to resistance to tamoxifen, which is presumed to be due to the ability of a breast cancer cell to bypass a response to normal endocrine therapies; this is more likely with co-expression of HER2 and HER3 [72]. Although there are currently no specific therapeutic targets to HER3, several studies indicate that activation of HER3 is a major cause of failure of anti-HER2 or anti-ER based therapies. Overexpression of HER3 has been shown to predict resistance to trastuzumab and lapatinib [73].

Prognostic Value

The prognostic value of HER3 is unknown. Studies have shown conflicting evidence and suggest that overexpression of HER3 may be associated with worse survival or of no prognostic use [74]. In a study of 177 primary breast cancers, low HER3 expression predicted recurrence in HER2 amplified breast cancer [74] and more aggressive clinicopatho-logical features.

Future Directions

There are a number of HER3-targeted biologic therapies in preclinical phases of study in breast, colorectal, lung and ovarian cancer. One example is Seribantumab (MM-121) [75], a monoclonal antibody that inhibits HER3 activation, which is in Phase II testing for preoperative triple negative breast cancer (TNBC) and HR-positive, HER2-negative breast cancer in combination with chemotherapy. However, concerns have been raised regarding intolerable side effects of HER3-targeted therapies, resulting in discontinuation of a previous study [72, 75, 76].

Human Epidermal Growth Factor Receptor 4 (HER4)

HER4 mediates both anti-proliferative and pro-apoptotic activity; its function is not yet fully understood [77, 78] and therefore, its overall expression in breast cancer remains unknown.

Association with Age

As with HER3, high expression has been noted in older women (≥70 years) with primary breast cancer, but no direct comparison to younger patients has been made [6].

Predictive Value

Recent studies have shown that expression of HER4 improves OS of breast cancer patients with ER-positive tumours, raising the possibility that HER4 influences response to endocrine therapy [79]. This has been shown in mouse models where HER4 suppression has been associated with tamoxifen resistance [80].

Prognostic Value

Expression of HER4 is associated with good prognosis, with better BCSS, in early breast cancer and correlates with ER-positivity and lower tumour grade [81-83].

Future Directions

There are few clinical trials to investigate anti-HER4 therapies, compared to anti-HER3 therapies. Anti-HER4 antibodies, such as clone P6-

1, have been shown to cause growth inhibition of breast cancer cells in vitro [73].

Figure 5 summarises therapeutic targets against the HER family, currently under investigation.

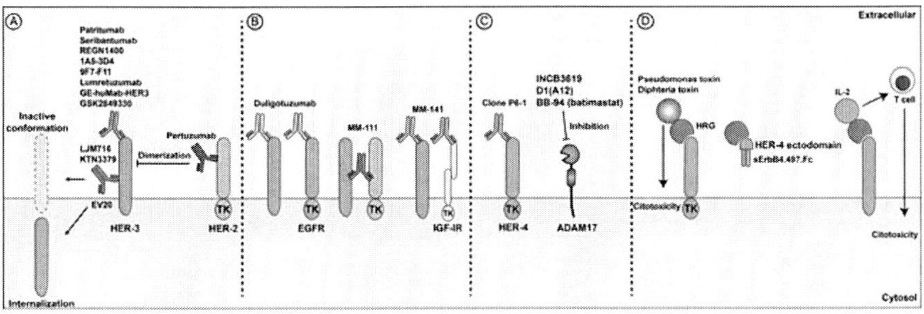

Figure 5. A summary of therapeutic targets including HER3, HER2, HER4 and EGFR (permission to reuse not required).

Vascular Endothelial Growth Factor (VEGF)

VEGF is a signalling protein expressed on vascular endothelial cells as well as carcinoma cells. Isoform VEGF-A has been shown to have an essential role in angiogenesis and tumour blood supply which is a major element in tumour growth and metastasis [84, 85] (Figure 6) and is overexpressed in approximately 72-98% of breast cancers [86].

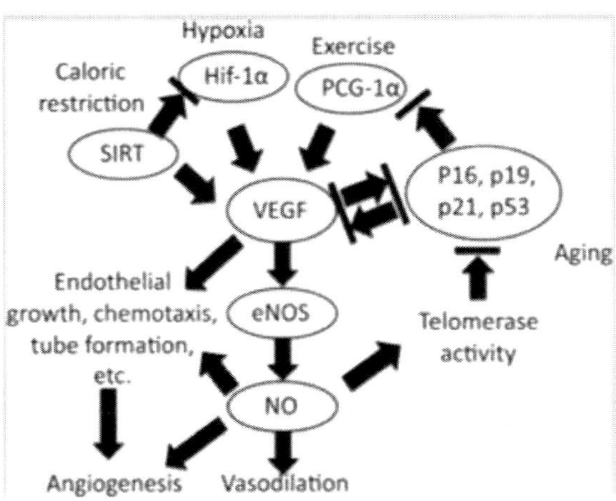

Figure 6. Overview of angiogenic pathways (permission to reuse not required).

Association with Age

Advancing age is associated with a decline in VEGF in normal tissue [87] Conversely, in a study of 575 surgical specimens from older women with primary breast cancer by Syed et al. [6], high levels were found in these patients.

Predictive Value

High VEGF expression has been linked to failure of tamoxifen therapy [88]. Foekens et al. [88] measured levels of VEGF in primary breast tumours from 845 patients who developed recurrence during follow-up. Patients with higher or intermediate levels of VEGF showed a poor rate of response to first-line tamoxifen therapy compared to those with low VEGF levels.

The benefits of anti-VEGF therapy, for example, bevacizumab (a monoclonal antibody) when combined with standard chemotherapy for first-line treatment of metastatic HER2-negative breast cancer have been confirmed in the Phase III RIBBON-1 trial (Regimens in Bevacizumab for Breast Oncology) [89]. The combination of bevacizumab with chemotherapy improved clinical benefit in terms of increased progression-free survival (PFS).Bevacizumab is approved for use in the US.

Prognostic Value

In the same study by Foekens et al. [88], higher levels of VEGF were associated with shorter PFS and OS. This has been confirmed in other breast cancer studies [86, 90].

Future Directions

Following RIBBON-1 [89], clinical trials of bevacizumab are ongoing. The current focus is on identifying subgroups of patients, based on expression of other biomarkers, who may benefit from this therapy [91].

Tumour Suppressor Genes

These are a group of genes prone to genetic change, which results in tumorigenesis. Tp53 mutations are estimated to occur in 50% of all cancers [92], however, a comprehensive meta-analysis revealed that only approximately 20% of all breast cancer cases express mutant p53 [93].

Tumour Suppressor Protein 53 (p53)

A mutation of the p53 gene results in the ability of damaged cells to induce apoptosis [94], resulting in cell death.

Association with Age

The likelihood of a p53 mutation increases with age. A study based on the GLOBOCAN and International Agency for Research on Cancer (IARC) databases showed that p53 mutations accounted for approximately 25% of the aging-related rise in breast cancer worldwide [95].

Predictive Value

The impact of Tp53 status on antihormone treatments has been investigated in two retrospective series, which provided evidence that Tp53 mutation may affect tamoxifen response [96].

Yamashita et al. [97] found that p53 protein accumulation was associated with greater likelihood of resistance to endocrine therapy [97] in metastatic breast cancer, as well as reduced post-relapse survival, compared to cases where p53 expression was low.

Prognostic Value

Low expression of mutant p53 has been linked to better DFS and OS in breast cancer [98]. These finding have been found to be similar across all age groups [6, 19].

Future Directions

Several retrospective studies have investigated a potential prognostic and therapeutic predictive role for mutant p53 in breast cancer, however with conflicting results [99]. Several compounds are being investigated which can reactivate the mutant p53 protein and convert it to a conforming type. Some of these compounds (e.g., PRIMA-1) have been found to exhibit anticancer activity in preclinical models of breast cancer [99].

Breast Cancer Gene (BRCA) 1 and 2

BRCA1 and 2 are genes that produce tumour suppressor proteins. Specific inherited mutations in BRCA1 and BRCA2 increase the risk of breast, ovarian and pancreatic cancers. BRCA mutations are the most common genetically inherited mutations which increase the risk of breast cancer; around 5% of breast cancers are caused by a BRCA mutation [24].

Associations with Age

Estimates from pooled data of 8,139 patients, of whom 500 had a germline mutation in BRCA1 and BRCA2, give an average cumulative risk of BRCA1 mutation carriers by age 70 years of 65% and 45% for BRCA 2. The risk of developing breast cancer with a BRCA mutation increased with age up to 70 years and then plateaued [100]. Although the relative risk of developing breast cancer is the same, there is a loss of expression of BRCA1 and BRCA2 in older women [6] compared to their younger counterparts.

Predictive Value

There is some laboratory and clinical data to suggest that BRCA1 associated breast cancer may be resistant to tamoxifen [101]. A retrospective study, comparing outcomes in early sage BRCA mutation and sporadic breast cancer treated with endocrine therapy, notes a lower OS in the BRCA carrier group [102].

Prognostic Value

In an analysis of 1940 breast cancer cases [22], altered BRCA1 was associated with shorter disease-free interval, as well as development of recurrence. A systematic review and meta-analysis by van den Broek et al. [103] examined the evidence for breast cancer prognosis of BRCA1/BRCA2 mutation carriers. They concluded that current evidence did not support worse breast cancer survival of mutation carriers. This is contradicted in a recent retrospective cohort study by Vocka et al. [37] that analysed the effect of clinicopathological features on prognosis in BRCA1/2 carriers. They found that mutation carriers with ER-positive disease were high-risk patients in terms of risk of breast cancer death and recurrence.

Future Directions

Poly (ADP-ribose) polymerases (PARPs) are important enzymes in DNA damage repair mechanisms, which BRCA1/2 mutations rely on

(Figure 7). The PARP inhibitor, olaparib, has been approved for use in HER2-negative metastatic breast cancer with an inherited BRCA1/2 mutation. The phase III OlympiAD trial has shown a PFS benefit of olaparib compared to single-agent chemotherapy in this setting, but no OS benefit [104]. There are currently over 30 actively recruiting trials involving olaparib to determine the exact benefit as well as comparison with other immunotherapy agents.

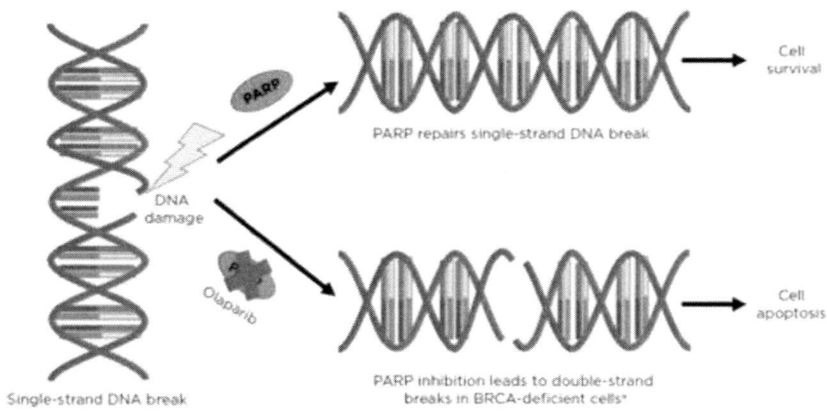

Figure 7. Mechanism of action of olaparib [104] (permission to reuse not required).

Phosphatase and Tensin Homolog (PTEN)

PTEN is a negative regulator of the PI3K/ALT signalling pathway (Figure 4) which is associated with tumorigenesis in multiple cancers including breast. Less than 5% of breast cancers occur because of a PTEN mutation, however, around 25% have significantly low levels of PTEN [105].

Association with Age

Association of PTEN with age is not clearly understood. Dean et al. [106] studied 101 breast cancers and found that PTEN loss was associated with younger age at the onset of breast cancer. Shoman et al. [107] found no association between PTEN expression and age (range 28-93 years) in 100 patients.

Predictive Value

PTEN expression is postulated to have an effect on acquired resistance to tamoxifen. Shoman et al. [107] found an association between downregulation of PTEN expression in ER-positive tumours and failure to respond to adjuvant tamoxifen.

Prognostic Value

Reduced PTEN expression has been associated with more advanced disease [107] and PTEN gene deletion has shown direct association with HER2 amplification and poorer prognosis in HER2 positive breast cancer overall [108].

A study of 49 primary breast cancer patients treated with adjuvant tamoxifen by Tanic et al. [109] found that the loss of PTEN expression was associated with shorter DFS, BCSS and OS, compared to high expression.

A meta-analysis by Li et al. [110] based on 27 studies involving 10,231 breast cancer patients found that PTEN loss was associated with worse DFS and OS, larger tumour size, lymph node metastasis and triple-negative phenotype.

Future Directions

Each stage of the downstream signalling pathway following on from PTEN (Figure 4) consists of potential therapeutic targets (Figure 4). Mammalian target of rapamycin (mTOR) is a serine/threonine protein kinase, downstream of AKT. Once activated it is responsible for limiting the proliferative signals transmitted by upstream effectors [111].

Everolimis is a rapamycin analogue that targets mTOR and was the first drug to be approved for the treatment of HR-positive/HER2-negative metastatic breast cancer which has progressed on first-line therapy with an aromatase inhibitor (AI) [112]. The Breast Cancer Trials of OraL EveROlimus-2 (BOLERO-2) trial [113] demonstrated that the addition of everolimus to exemestane (a steroidal AI) markedly prolonged PFS in HR positive, advanced breast cancer with disease recurrence following prior non-steroidal AIs (Figure 8).

An invitro study by Owusu-Brackett et al. [114] tested TAK228 (mTOR inhibitor). TAK228 enhanced the efficacy of eribulin (a chemotherapy agent) in TNBC, but further investigation is require to determine patient selection for this combination therapy.

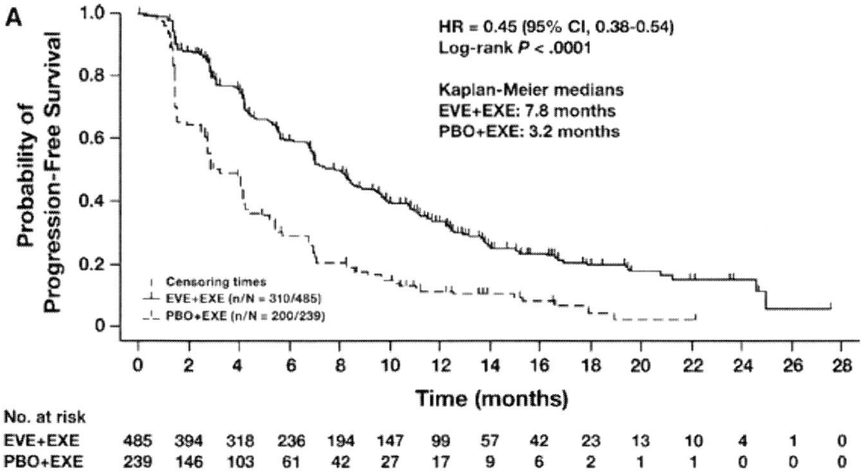

Figure 8. Kaplan-Meier estimates of progression-free survival of patients treated with everolimus plus exemestane versus exemestane alone. *CI* confidence interval, *HR* hazard ratio, *EVE* everolimus, *EXE* exemestane, *PBO* placebo [113] (permission to reuse not required).

Liver Kinase B1 (LKB1)

LKB1 is a serine-threonine kinase described in the development of Peutz Jagher's syndrome where around 45% of patients develop breast cancer in their lifetime [14], among other cancers.

Association with Age

High expression of LKB1 has been noted in older women (>70 years) compared to younger patients [6, 14].

Predictive Value

There is little in the present literature about the predictive value of LKB1 in breast cancer. In a study of older breast cancer patients undergoing surgery [14], LKB1 expression was associated with better survival outcome among patients receiving adjuvant endocrine therapy.

Prognostic Value

In the same study [14], within the older age group, LKB1 expression was associated with poor prognostic factors including high tumour grade and high expression of Ki-67 and HER2.

Contrary to this, Chen et al. [115] found that high LKB1 expression was predictive of better survival in patients with HER2-positive tumours.

Future Directions

Future studies are required to investigate and compare the expression of LKB1 in the younger compared to older patient population and delineate the precise role of LKB1 as a therapeutic target. Metformin has been shown to selectively inhibit proliferation of LKB1 positive cancers including breast cancer [116], which needs further evaluation.

Cell Proliferative Markers

Ki-67

Ki-67 is a protein that is associated with cell proliferation. It is difficult to determine the overall high expression of Ki-67 in breast cancer due to the lack of consensus over the method of detection and definition of 'high' expression [117, 118].

Association with Age

In a cohort of 462 patients with stage I-III breast cancer [119], younger age (≤40 years) was associated with high expression of Ki-67. In a cohort of 575 older women (≥70 years), low expression of Ki-67 was seen [6]. These findings suggest a less proliferative disease in older women.

Predictive Value

Measurement of Ki-67 has been utilised to monitor the tumour proliferation index in neoadjuvant settings, particularly for chemotherapy treatment. A high Ki-67 proliferation index is an independent predictor factor for pCR after neoadjuvant chemotherapy (NACT) [120].

The IMPACT trial was a neoadjuvant randomised trial evaluating adjuvant treatment with anastrozole, tamoxifen or in combination, in 330 postmenopausal patients with ER-positive breast cancer [121]. A secondary objective of the study was to evaluate changes in Ki-67 expression after two weeks of treatment. Higher expression of Ki-67 after two of endocrine therapy was associated with shorter recurrence free survival (RFS), larger tumour size at baseline and lower ER level.

Prognostic Value

A meta-analysis by Petrelli et al. [122] exploring the prognostic value of Ki-67 in early breast cancer analysed data of 64,196 patients from 41 studies. Ki-67 was an independent prognostic indicator in terms of OS and expression of >25% was associated with greater risk of death compared with lower expression rates.

A meta-analysis by de Azambuja et al. [123] identified 12,155 patients from 46 studies where Ki-67 was measured. Ki-67 positivity was associated with higher rate of relapse and worse survival in patients with early breast cancer.

Future Directions

Despite endorsements by several international guidelines, measurement of Ki-67 is yet to gain widespread application as a prognostic or predictive marker, which is mainly to do with wide variation in methodology of detection and lack of standardisation [118].

Recent studies in renal cancer have focused on Ki-67 targeted therapy but this is not yet done in breast cancer [124]; therapies targeting Ki-67 in the setting of renal carcinoma are in the preclinical stages.

Anti-Apoptosis Markers

B Cell Lymphoma 2 (BCL-2)

BCL-2 gene is the founding member of the BCL-2 family of regulator proteins that regulate cell death by either inhibiting or inducing apoptosis.

In a retrospective study of 605 cases of breast cancer, 53.8% showed BCL-2 expression [125].

Association with Age

In the same study [125], BCL-2 positive expression was associated with young age (<50 years). High expression of BCL-2 has been found in a cohort of women ≥70 years with primary breast cancer treated by surgery [6], compared to younger patients (<70 years) and also in a series of older women with TNBC [13] (≥70 years). Biological characterisation of 14,007 primary breast cancers by Daidone et al. [43] showed an increase in BCL-2 expression with age (≥65 years compared to <65 years).

Predictive Value

High BCL-2 has been linked to better clinical response to tamoxifen in ER-positive metastatic breast cancer [126].

Tamoxifen induces apoptosis in breast cancer cells. BCL-2 genes interfere with apoptosis in various ways. A study by Zhang et al. [127] investigated the competing effects of tamoxifen and BCL-2 gene products and found that tamoxifen induced down-regulation of BCL-2, thus inducing apoptosis.

Prognostic Value

A meta-analysis by Callagy et al. [128] reviewed 5,892 cases of female breast cancer where BCL-2 had been measured. They concluded that patients with high expression of BCL-2 had better OS and DFS. This has been reproduced in other studies [129, 130].

Specifically in older women, BCL-2 has been shown to correlate with better DFS and BCSS [6, 131].

Future Directions

Therapeutic compounds have been developed to attempt to disrupt the action of BCL-2, known as BH3-mimentics (Figure 9). These BCL-2 inhibitors have shown ability to restrict tumour growth in xenograft models [132] and there are a number currently under further investigation in the clinical trial setting for example, Navitoclax [133].

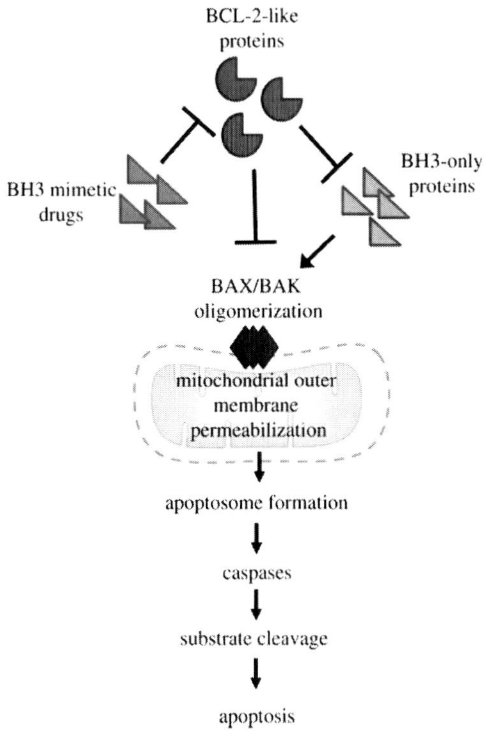

Figure 9. Action of BH3 mimetic drugs [132] (permission to reuse not required).

Cell Differentiation Markers

Cytokeratins (CKs)

Breast cancer originates from basal (myoepithelial) or luminal (glandular) tissue that express specific cytokeratins. In normal breast tissue, CK 5, CK 5/6and CK14 are expressed in myoepithelial cells and CK7, CK8, CK18, CK19, CK20 in ductal epithelium [134].

Association with Age

High expression of CK5/6, CK5, CK18 and low expression of CK7/8 has been seen in the older population of primary breast cancer patients [6], compared to younger counterparts (<70 years). In a cohort of patients with TNBC, CK7/8 positivity was lower and CK18 higher, compared to a comparative younger series [135]. A meta-analysis to determine the clinicopathological significance of CK18 in breast cancer, confirmed higher expression in the older population ≥50 years, compared to <50 years [136].

Predictive Value

There is minimal evidence in the literature to support the predictive values of cytokeratins, which encompass a large heterogeneous group of markers. In a study of CK20 in a human cell line, CK20 expression enhanced the metastatic potential of breast cancer cells and was thought to be a potential therapeutic target in tamoxifen-resistant breast cancer [137].

An experimental study by Bozionellou et al. [138] has shown that chemotherapy resistant CK19 mRNA-positive cells in the peripheral blood and bone marrow can be effectively targeted by trastuzumab administration.

Prognostic Value

A study of 611 breast cancers by van de Rijn et al. [139] found that expression of CK 17 and CK 5/6 in tumour cells was associated with poor clinical outcome. Multivariate analysis showed that in node-negative cancer, expression of CK 17 and C5/6 was a prognostic factor independent of size and grade.

A literature review by Haupt et al. [140] looked at the literature relating to basal-like breast carcinomas and confirmed that expression of basal CKs was an independent prognostic factor in lymph node negative cases, but in lymph node positive cases this was unclear. Overall, the evidence was suggestive that basal-like breast carcinomas are associated with worse clinical outcome.

Future Directions

Cytokine expression has been found to be associated with response to PET in a small cohort of older women with primary operable breast cancer [15] and these findings need validating in a larger cohort.

OTHERS

Mucin 1 (MUC1)

MUC 1 is an epithelial cell surface protein that is aberrantly overexpressed in over 90% of breast cancer as well as other cancers [141].

Association with Age

There is little in the current literature to describe the association of the expression of MUC1 with age in breast cancer. A study in older compared to younger women having surgery did not show any statistically significant difference in expression [6].

Predictive Value

In ER-positive patients, increased expression of MUC1 is associated with tamoxifen resistance [142].

Prognostic Value

Breast cancers which exhibit increased and mislocalised expression of MUC1 are more likely to metastasise [142].

High levels of MUC1 have been associated with poor clinical outcome [143-145]. Jing et al. [143] accessed expression levels of MUC1 in breast cancer and normal tissues from the Oncomine database. Abnormally high levels of MUC1 was associated with poor prognosis, with reduced FRS and disease-specific survival (DSS).

An analysis of 171 cases of invasive breast cancer by McGuckin et al. [144] showed patients whose tumours had high expression of MUC1 has significantly poorer DFS and OS, when compared with those with lower expression.

Future Directions

MUC1 targeted therapies have been in development for the last 30 years, however effective benefit in clinical trials has not yet been achieved [146].

A monoclonal antibody GGSK-1/30 targeting human tumour-associated MUC 1 has shown promising initial results in human breast tissue analysis at determining the severity of disease [141].

Amplified in Breast Cancer 1 (AIB1)

AIB1 is a member of the steroid receptor coactivator family and mediates the activities of nuclear receptors including ER and PgR. AIB1 is overexpressed in 30-60% of breast cancers [147].

Association with Age

AIB1 expression was measured in 185 breast cancer surgical resection specimens; there was no difference noted between expression and age (≤50 years compared to >50 years) [148].

Predictive Value

In postmenopausal women with primary breast cancer, high AIB1 has been implicated as a marker of good response to tamoxifen treatment as adjuvant therapy. Furthermore, it was a prognostic marker of decreased RFS in patients who did not receive adjuvant systemic therapy [149].

Osborne et al. [150] measured AIB1 and HER2 levels in 316 breast cancer patients. In patients who received tamoxifen, high AIB1 expression was associated with worse DFS, indicative of tamoxifen resistance. Patients whose tumours expressed AIB1 and HER2 had worse outcomes with tamoxifen therapy than other combinations.

These two studies are conflicting in their findings.

Prognostic Value

The International Breast Cancer Study Group's trial BIG 1-98 randomised 1,396 Danish patients to adjuvant tamoxifen and/or AIs [151]. AIB1 expression correlated to expression of HER2 and high grade as well as poor DFS and OS.

In a study of 185 breast cancers by Lee at al [148], high AIB1 expression was associated with lower DFS and OS. These findings were most significant in the ER-negative cohort.

Future Directions

Steroid receptor coactivators such as AIB1 lack a high affinity ligand binding domain and therefore are considered difficult drug targets [152].

An AIB1 small molecule inhibitor, bufalin, has been shown to effectively inhibit tumour growth in TNBC in a mouse model [153], suggesting a potential target for TNBC treatment. Bufalin and other small molecule inhibitors are still in pre-clinical phases of investigation [152].

Other Potential Biomarkers

It is not possible to discuss every relevant biomarker in breast cancer in this chapter, of which are there many hundreds. We have described so far a number of biomarkers which are known to the authors to be relevant in ER-positive breast cancer and which are under further investigation in this setting.

Some more recent biomarkers which have been discovered to have a prognostic role in breast cancer and involvement in acquired endocrine resistance include glutaminases [18], zinc transporters [154], Cathepsin [155], cell division regulators (such as CDC20) [17] and other members of the nuclear receptor superfamily (such as RXRG) [156].

With such a plethora of literature of copious numbers of biomarkers, it can be daunting to understand how we should use this information in clinical practice. In the next section of the chapter, we will discuss high throughput measurement of biomarkers in the research and clinical settings.

High Throughput Measurement of Biomarkers

The development of high-throughput technologies to investigate genetic, epigenetic and proteomic changes has helped to unravel the complexity of breast cancer biology [157]. The realisation of breast tumour heterogeneity prompted the development of a molecular classification system, which constituted the first step towards the establishment of personalised medicine in breast cancer.

Concept of Tissue Microarray (TMA)

The majority of studies to date that have profiled the biology of breast cancer have used (SE specimens to do so and analyse individual patient specimens one at a time.

This creates potential bias in having not included the large group of patients who do not have surgery and furthermore, cannot provide any insight when considering neoadjuvant therapy. Performing IHC analysis on patient specimens one at a time is costly and time-consuming.

As an alternative to SE profiling, CNB, which is usually obtained at diagnosis in breast cancer patients, should be considered for the study of tumour biology. CNB samples can be obtained from all older patients diagnosed with breast cancer regardless of primary treatment. Another advantage of CNB compared to SE is that preservation of CNB tissue leads to faster penetration of the tissue by a fixative agent, resulting in less chance of enzyme degradation and thus, better preservation of biological features [158, 159].

The technique of TMA for use in breast cancer was first reported by Kononen et al. [160] in 1998. The construction of TMAs involves embedding multiple fragments of tumour tissue in a single paraffin block for the purpose of high throughput analysis. In addition to maximising tissue resources, TMA has the advantage of facilitating analysis and evaluation of tissue-based assays in an efficient, cost-effective and uniform condition [161]. Tissue micro-array has become standard technique to examine tissue biology in detail but primarily in SE specimens. There are many challenges faced when manipulating TMA technique to utilise CNB for profiling tumour biology, including complex construction given the small diameter of the biopsy and erosion of biopsy after sectioning for initial diagnosis. After a comprehensive systematic literature review on the subject [16], our team has successfully managed to develop an optimal technique to construct TMAs from CNB samples [15]; this gives us the unique opportunity to utilise CNB for profiling tumour biology.

We therefore suggest that CNB TMA is the optimum method for studies assessing biology in breast cancer.

Prediction Tools Available

There are a number of prediction tools currently available on the market, which utilise either measurement of tissue protein or gene assay.

Tools Based on Tissue Protein Analysis

PREDICT [9] is a validated tool to predict overall and BCSS survival for women treated with early breast cancer in the UK. It considers the following factors: age, menopausal status, ER status, HER2 status, Ki-67 status, tumour size, tumour grade, number of positive nodes. Using the Eastern Cancer Registration and Information Centre dataset, information was collated for 5,694 women who had surgery for invasive breast cancer. Breast cancer mortality models were derived from this data using the PREDICT tool and validated against an external dataset. Model calibration was good for both data sets.

Adjuvant! Online [8] was developed in patients aged ≤69 years and aids decision making regarding adjuvant therapies. The tool considers the following information: age, menopausal status, comorbidity estimate, tumour size, number of positive axillary nodes, ER status. Outcomes for OS and DFS seen in clinical trials are reasonably modelled by Adjuvant Online! in the younger age group [8], but not for patients ages ≥65 years [162].

These tools have been developed and validated based on a collection of clinical and pathological information from large patient populations, not measurement of protein-based biomarkers from individual patients.

Tools Based on Gene Assay

There are a number Of genomic tests which analyses the activity of certain genes in breast cancer, these include Oncotype DX [10], Mammaprint [163] and EndoPredict [164]. These tools are based on personalised measurement of biomarkers in the individual patient's tumour sample.

Only Oncotype DX has been recommended for use by NICE in the UK [165]. Oncotype DX is a 21-gene assay to assist in decision making regarding adjuvant chemotherapy in patients with ER positive and HER2 negative tumours. In has been validated in a prospective trial setting [10].

FUTURE DIRECTIONS SUMMARY

There is a potential for extensive biomarker research, as we have described in this chapter, to be used in detection, treatment and measurement of recurrence risk of breast cancer. Current tools are not validated for used in older women (>70 years) and are primarily concerning adjuvant treatment. Once we have a better understanding of the complex roles and interactions of biomarkers in breast cancer there is potential for prediction tools to be employed at all stages of the treatment pathway.

In order to produce a tool which could be generalised to all patients with breast cancer, we need consensus regarding how to measure currently undefined biomarkers, such as Ki-67 and uniform agreement on definitions of high expression. Despite the ageing population and increased likelihood of ER-positive breast cancer in older women, there is a lack of research focusing on the biology of breast cancer in this cohort of patients. Increasing ER-positivity in this group is of importance at all stages of the treatment patients ((neo)adjuvant, primary and advanced) as they are more likely to consider less aggressive alternative therapies compared to surgery and chemotherapy.

In the future, optimal personalised treatment for the individual patient with breast cancer based on the unique biological characteristics of that cancer is achievable.

CONCLUSION

ER, PgR and HER2 are biomarkers routinely measured by IHC on biopsy samples taken at diagnosis for breast cancer. Many other biomarkers are associated with predicting prognosis and response to therapy in breast cancer, but are not currently measured outside of a research setting. Differences in expression of biomarkers between younger and older women has been confirmed for ER, HER2, Tp53 and cytokeratins. The mechanism of endocrine resistance is complex and biomarkers found to be associated with this are HER3, HER4, EGFR, Tp53, BRCA1, PTEN, BCL-2, MUC1 and AIB1. The biomarkers ER, PgR, HER2, HER4, EGFR, VEGF, Tp53, PTEN, Ki-67, BCL-2, cytokeratins, MUC1 and AIB1 have been shown to have an association with survival. Once we have a better understanding of the complex roles and interactions of biomarkers in breast cancer, there is potential for prediction tools to be developed and employed at all stages of the treatment pathway [166].

REFERENCES

[1] Puntmann VO. *Postgrad. Med. J.* 2009; 85, 538.

[2] Anderson WF, et al. *J. Natl. Cancer Inst.* 2014; 106, dju165.

[3] Burstein MD, et al. *Clin. Cancer Res.* 2015; 21, 1688-1698.

[4] Lehmann BD, et al. *J Clin Invest.* 2011; 121, 2750-2767.

[5] Yersal O, et al. *World J Clin Oncol.* 2014; 5, 412-424.

[6] Syed BM, et al. *Br. J. Cancer.* 2013; 108, 1042-1051.

[7] Dai XC, et al. *J. Cancer.* 2017; 8, 3131-3141.

[8] Ravdin PM, et al. *J. Clin. Oncol.* 2001; 19, 980-991.

[9] Wishart GC, et al. *Breast Cancer Res.* 2010; 12, R1.

[10] Sparano JA, et al. *N. Engl. J. Med.* 2015; 373, 2005-2014.

[11] Turner NC, et al. *Lancet.* 2017; 389, 2403-2414.

[12] Lumachi F, et al. *World J Biol Chem.* 2015; 6, 231-239.

[13] Syed BM, et al. *PloS one.* 2014; 9(, e100573.

[14] Syed BM, et al. *Cancers.* 2019; 11, 149.

[15] Albanghali M. *PhD Thesis,* University of Nottingham. November 2016.
[16] Albanghali M, et al. *Histopathology.* 2015; 68, 323-332.
[17] Alfarsi LH, et al. *Breast Cancer Res. Treat.* 2019; 178, 535-544.
[18] El-Ansari R, et al. *Breast Cancer Res. Treat.* 2019; 175, 27-38.
[19] Cheung KL, et al. *Crit. Rev. Oncol. Hematol.* 2008; 67, 263-267.
[20] Syed BM, et al. *Br. J. Cancer.* 2011; 104, 1393-1400.
[21] Abd El-Rehim DM, et al. *Int. J. Cancer.* 2005; 116, 340-350.
[22] Rakha EA, et al. *Hum. Pathol.* 2008; 39, 857-865.
[23] Aleskandarany MA, et al. *Breast Cancer Res. Treat.* 2011; 127, 591-599.
[24] UK CR. *Breast cancer survival.* 2010-11.
[25] Kerlikowske K, et al. *J. Natl. Cancer Inst.* 2016; 109.
[26] Rhodes A, et al. *J Clin Pathol.* 2000; 53, 688-696.
[27] Waters EA, et al. *Breast Cancer Res. Treat.* 2012; 134, 875-880.
[28] Fisher B, et al. *J. Natl. Cancer Inst.* 1998; 90, 1371-88.
[29] Eiermann W. et al. *Ann. Oncol.* 2001; 12, 1527-1532.
[30] Preece PE, et al. *Br Med J* (Clin Res Ed). 1982; 284, 869-870.
[31] Caudle AS, et al. *Ann. Surg. Oncol.* 2011; 18, 932-938.
[32] Hind D, et al. *Database Syst. Rev.* 2006.
[33] Monypenny I. *UK Symptomatic Breast Audit* 1.4.2001-31.3.2002. British Association of Surgical Oncology. 2003.
[34] Early Breast Cancer Trialists' Collaborative G. *Lancet.* 2011; 378, 771-784.
[35] Sopik V, et al. *Breast Cancer Res. Treat.* 2017; 165, 391-402.
[36] Putti TC, et al. *Mod. Pathol.* 2005; 18, 26-35.
[37] Vocka M, et al. *Cancers.* 2019; 11, 738.
[38] NICE. Abemaciclib with an aromatase inhibitor for previously untreated, hormone receptor-positive, HER2-negative, locally advanced or metastatic breast cancer. *Nice Guideline.* 2019.
[39] Pernas S, et al. *Ther Adv Med Oncol.* 2018; 10, 1758835918786451.
[40] Dai X, et al. *J Cancer.* 2016; 7, 1281-1294.
[41] Thakkar JP, et al. *Oncologist.* 2011; 16, 276-285.
[42] Clark GM, et al. *J Clin Oncol.* 1984; 2, 1102-1109.

[43] Daidone MG, et al. *Crit Rev Oncol Hematol.* 2003; 45, 313-325.
[44] Nordenskjöld A, et al. *Breast Cancer Res Treat.* 2016; 160, 313-322.
[45] Dowsett M, et al. *J Clin Oncol.* 2005; 23, 7512-7517.
[46] Park S, et al. *Ann Surg Oncol.* 2013; 20, 1505-1513.
[47] Cancello G, et al. *Ann Oncol.* 2013; 24, 661-668.
[48] Van Asten K, et al. *Oncologist.* 2019; 24, 165-171.
[49] Wieduwilt MJ, et al. *Cell Mol Life Sci.* 2008; 65, 1566-1584.
[50] Ayoub NM, et al. *Breast Cancer (Dove Med Press).* 2019; 11, 53-69.
[51] Asati V, et al. *Eur J Med Chem.* 2016; 109, 314-441.
[52] Rimawi MF, et al. *Cancer.* 2010; 116, 1234-1242.
[53] Osborne CK, et al. *Annu Rev Med.* 2011; 62, 233-247.
[54] Diab SG ER, et al. *J Natl Cancer Inst.* 2000; 92, 550-556.
[55] Tsutsui S, et al. *Clin. Cancer Res.* 2002; 8, 3454.
[56] Gasparini G. *Oncologist.* 2000; 5, 37-44.
[57] Bernsdorf M, et al. *Breast Cancer Res. Treat.* 2011; 126, 463-470.
[58] Johnston S, et al. *J Clin Oncol.* 2008; 26, 1066-1072.
[59] Nakai K, et al. *Am J Cancer Res.* 2016; 6, 1609-1623.
[60] Soto-Perez-De-Celis E, et al. *Expert opin inv drug.* 2018; 27, 787-801.
[61] Support MC. HER2 positive breast cancer. *Macmillan Cancer Information and support.* 2016.
[62] Pinhel I, et al. *Breast Cancer Res.* 2012; 14, R46.
[63] Laird-Fick HS, et al. *J Geriatr Oncol.* 2013; 4, 362-367.
[64] Cameron D, et al. *Lancet.* 2017; 389, 1195-1205.
[65] (NICE) NIfHaCE. Early and locally advanced breast cancer: diagnosis and management. *Nice Guideline*. 2018; 101.
[66] Cianfrocca M, et al. *Oncologist.* 2004; 9, 606-616.
[67] Excellence NIfHaC. Pertuzumab for adjuvant treatment of HER2-postive early stage breast cancer. *Technology appraisal guidance* [TA569]. 2019.
[68] Diéras V, et al. *Lancet Oncol.* 2017; 18, 732-742.
[69] Slamon DJ, et al. *Science.* 1987; 235, 177.
[70] Travis A, et al. C-erbB-3 in human breast carcinoma: expression and relation to prognosis and established prognostic indicators. *Br J Cancer.* 1996;74(2):229-33.

[71] Li Q, et al. *Oncotarget.* 2017; 8, 67140-67151.
[72] Mishra R, et al. *Oncol Rev.* 2018; 12, 355.
[73] Mota JM, et al. *Oncotarget.* 2017; 8, 89284-89306.
[74] Luhtala S, et al. *BMC Cancer.* 2018; 18, 1045.
[75] Merrimack. A trial of preoperative MM-121 with paclitaxel in HER2-negative breast cancer. *ClinicalTrialsgov.* 2016; 2016, NCT01421472.
[76] Sequist LV, et al. *Oncologist.* 2019; 24, 1095-1102.
[77] Naresh A, et al. *Cancer Res.* 2006; 66, 6412.
[78] Sartor CI, et al. *Mol Cell Biol.* 2001; 21, 4265-4275.
[79] Witton CJ, et al. *J. Pathol.* 2003; 200, 290-297.
[80] Naresh A, et al. *Cancer Res.* 2008; 68, 6387-6395.
[81] Koutras AK, et al. B*r. J. Cancer.* 2008; 99, 1775-1785.
[82] Suo Z, et al. *J Pathol.* 2002; 196, 17-25.
[83] Kew TY, et al. *Br. J. Cancer.* 2000; 82, 1163-1170.
[84] Folkman J. Role of angiogenesis in tumor growth and metastasis. *Seminars in Oncology.* 2002; 29(6, Supplement 16), 15-18.
[85] Ferrara N. *Arter. Thromb. Vasc. Biol.* 2009; 29, 789-791.
[86] Liu Y, et al. *Breast Cancer Res Treat.* 2011; 129, 175-184.
[87] Lähteenvuo J, et al. *Circ Res.* 2012; 110, 1252-1264.
[88] Foekens JA, et al. *Cancer Res.* 2001; 61, 5407.
[89] Robert NJ, et al. *J Clin Oncol.* 2011; 29, 1252-1260.
[90] Gasparini G. *Oncologist.* 2000; 5, 8.
[91] Roviello G, et al. *Eur J Cancer.* 2017; 75, 245-258.
[92] Gasco M, et al. *Breast Cancer Res.* 2002; 4, 70-76.
[93] Pharoah PD, et al. *Br. J. Cancer.* 1999; 80, 1968-1973.
[94] Lowe SW, et al. *Science.* 1994; 266, 807.
[95] Richardson RB. *Cell Cycle.* 2013; 12, 2468-2478.
[96] Fernandez-Cuesta L, et al. *Int J Cancer.* 2011; 128, 1813-1821.
[97] Yamashita H, et al. *Breast Cancer Res.* 2006; 8, R48.
[98] Thor AD, et al. *J Natl Cancer Inst.* 1992; 84, 845-855.
[99] Duffy MJ, et al. *Breast Cancer Res Treat.* 2018; 170, 213-219.
[100] Antoniou A, et al. *Am J Hum Genet.* 2003; 72, 1117-1130.
[101] Zhu Y, et al. *Nat Commun.* 2018; 9, 1595.
[102] Wesolowski R, et al. *J Clin Oncol.* 2009; 27, e22065.

[103] van den Broek AJ, et al. *PloS one.* 2015; 10, e0120189.
[104] Caulfield SE, et al. *J Adv Pract Oncol.* 2019; 10, 167-174.
[105] Saal LH, et al. *Nat Genet.* 2008; 40, 102-107.
[106] Dean SJR, et al. *Am J Clin Pathol.* 2014; 141, 323-333.
[107] Shoman N, et al. *Mod Pathol.* 2005; 18, 250-259.
[108] Lebok P, et al. *BMC Cancer.* 2015; 15, 963.
[109] Tanic N, et al. *Cancer Biol Ther.* 2012; 13, 1165-1174.
[110] Li S, et al. *Oncotarget.* 2017; 8, 32043-32054.
[111] Presti D, et al. *Cancers.* 2019; 11, 1242.
[112] NICE. Everolimus with exemestane for treating advanced breast cancer after endocrine therapy. *NICE Technology appraisal guidance.* 2016.
[113] Yardley DA, et al. *Adv Ther.* 2013; 30, 870-884.
[114] Owusu-Brackett N, et al. *Oncotarget.* 2019; 10, 5011-5019.
[115] Chen IC, et al. *Sci Rep.* 2016; 6, 21374.
[116] Guo Q, et al. *Mol Med Rep.* 2016; 13, 2590-2596.
[117] Brown JR, et al. *Lab Invest.* 2014; 94, 98-106.
[118] Abubakar M, et al. *Breast Cancer Res.* 2016; 18, 104.
[119] Alco G, et al. *Oncol Lett.* 2015; 9, 1046-1054.
[120] Alba E, et al. *Oncologist.* 2016; 21, 150-155.
[121] Smith IE, et al. *J Clin Oncol.* 2005; 23, 5108-5116.
[122] Petrelli F, et al. *Breast Cancer Res Treat.* 2015; 153, 477-491.
[123] de Azambuja E, et al. *Br. J. Cancer.* 2007; 96, 1504-1513.
[124] Menon SS, et al. *Clin Chim Acta.* 2019;491:39-45.
[125] Eom YH, et al. *J Breast Cancer.* 2016; 19, 252-260.
[126] Elledge RM, et al. *J Clin Oncol.* 1997; 15, 1916-1922.
[127] Zhang G-J, et al. *Clin Cancer Res.* 1999; 5, 2971.
[128] Callagy GM, et al. *BMC Cancer.* 2008; 8, 153.
[129] Cory S, et al. *Oncogene.* 2003; 22, 8590-8607.
[130] Dawson SJ, et al. *Br. J. Cancer.* 2010; 103, 668-675.
[131] Daidone MG, et al. *Crit Rev Oncol Hematol.* 2003; 45, 313-325.
[132] Campbell KJ, et al. *Open Biol.* 2018; 8, 180002.
[133] Chen J, et al. *Mol Cancer Ther.* 2011; 10, 2340.

[134] Shao MM, et al. Keratin expression in breast cancers. *Virchows Arch.* 2012; 461, 313-322.
[135] Syed BM, et al. *PLoS One.* 2014; 9, e100573.
[136] Yang J, et al. *Biosci Rep.* 2018; 38, BSR20171145.
[137] Min YS, et al. *Anticancer Res.* 2012; 32, 1221-1228.
[138] Bozionellou V, et al. *Clin Cancer Res.* 2004; 10, 8185.
[139] van de Rijn M, et al. *Am J Pathol.* 2002; 161, 1991-1996.
[140] Haupt B, et al. *Arch Pathol Lab Med.* 2010; 134, 130-133.
[141] Stergiou N, et al. *Int J Med Sci.* 2019; 16, 1188-1198.
[142] Haddon L, et al. *Clin Exp Metastasis.* 2015; 32, 393-403.
[143] Jing X, et al. *Oncol Rep.* 2019; 41, 801-810.
[144] McGuckin MA, et al. *Hum Pathol.* 1995; 26, 432-439.
[145] Maeda T, et al. *Cancer Res.* 2018; 78, 205.
[146] Taylor-Papadimitriou J, et al. *Biochem Soc Trans.* 2018; 46, 659-668.
[147] Chang AK, et al. Oncol Lett. 2012; 4, 588-594.
[148] Lee K, et al. *World J Surg Oncol.* 2011; 9, 139.
[149] Weiner M, et al. *Ann Oncol.* 2013; 24, 1994-1999.
[150] Osborne CK, et al. *J Natl Cancer Inst.* 2003; 95, 353-361.
[151] Alkner S, et al. *Breast Cancer Res Treat.* 2017; 166, 481-490.
[152] Lonard DM, et al. *Clin Cancer Res.* 2016; 22, 5403-5407.
[153] Song X, et al. *PloS one.* 2015; 10, e0140011.
[154] Ziliotto S, et al. Metallomics. 2019; 11, 1579-1592.
[155] Wilkinson RDA, et al. *J Oncol.* 2019; 2019, 3980273.
[156] Joseph C, et al. *Br. J. Cancer.* 2019; 121, 776-785.
[157] Dawson S-J, et al. *EMBO J.* 2013; 32, 617-628.
[158] Srinivasan M, et al. *Am J Pathol.* 2002; 161, 1961-1971.
[159] Pekmezci M, et al. *Patholog Res Int.* 2012; 2012, 917011.
[160] Kononen J BL, et al. *Nat Med.* 1998; 4, 844-847.
[161] Rimm DL, et al. Cancer J. 2001; 7, 24-31.
[162] de Glas NA, et al. *Lancet Oncol.* 2014; 15, 722-729.
[163] van de Vijver MJ, et al. *N Engl J Med.* 2002; 347, 1999-2009.
[164] Filipits M, et al. *Clin Cancer Res.* 2011; 17, 6012.

[165] Excellence NIfHaC. *Tumour profiling tests to guide adjuvant chemotherapy decisions in early breast cancer*. Diagnostics guidance DG34. 2018.

[166] Olivares-Urbano MA, et al. *J Cell Mol Med* 2020; 24, 139-148.

In: Estrogen Receptors
Editors: George G Chen et al.
ISBN: 978-1-53618-228-6

Chapter 4

ESTROGEN RECEPTOR ALPHA: STRUCTURE, FUNCTIONS AND CLINICAL ASPECTS IN ENDOCRINE SYSTEM

Xuemin Zhao[1] and Jing Li[2,*], PhD, MD
[1]Department of Internal Medicine,
Chengde Medical University, Chengde, P.R. China
[2]Department of Endocrinology and Metabolism,
Institute of Endocrinology,
Liaoning Provincial Key Laboratory of Endocrine Diseases,
The First Affiliated Hospital of China Medical University,
Shenyang, P.R. China

ABSTRACT

Estrogen receptors consist of nuclear estrogen receptors (ERα and ERβ) and membrane estrogen receptors. The structure of ERα is composed of five domains including A/B, C, D, E, F-domain. ERα distributed in several endocrine tissues and not only mediated the regulation of hypothalamus-pituitary-gonad axis but also function of other endocrine tissues. In this chapter, we would focus on the influences of estrogens on

* Corresponding Author's Email: jli23@cmu.edu.cn /lijingendocrine@126.com.

other endocrine tissues than hypothalamus-pituitary-gonad axis, especially those mediated by ERα. ERα activation may affect serum levels of thyroid hormones, especially under high estrogen conditions. ERα mediates the stimulatory actions of thyroid neoplasms development by estrogens. Through the ERs, estrogens can also regulate both immunocytes of innate immune system and those of adaptive immune system. In addition, ERα may mediate protective actions on the development of type 2 diabetes, and lack of estrogens may promote insulin resistance and hyperglycemia. ER subtype-specific agonist or antagonist may become new strategies for the treatment of estrogen-dependent endocrine and metabolic disorders.

Keywords: Estrogen receptor alpha, thyroid neoplasms, pancreatic islet, glucose metabolism

TYPES OF ESTROGEN RECEPTORS

Estrogen receptors (ERs) consist of nuclear estrogen receptors (nER) which are members of transcription factors, and membrane estrogen receptors (mER). mER also further comprised of G protein-coupled estrogen receptor-1 (GPER1, also named GPR30) and another membrane-bound ER (referred as ER-X).

ER alpha (ERα, also known as ESR1) is encoded by the gene on chromosome 6 (6q25.1) and distributed in many organs, such as uterus and pituitary gland, breast, liver, kidney, white adipose tissue, prostate, ovary, testes, skeleton, brain, and immune system [1]. ER beta (ERβ, also known as ESR2), product of genes on chromosome 14 (14q22-24) [2-3], have widely been found in granulosa cells of the ovary, colon, salivary gland, vascular endothelium, lung, bladder, prostate, testes, skeleton, brain, and immune system. GPER1, product of genes on chromosome 7, is widely present in central and peripheral nervous system, uterus, ovaries, mammary glands, testes, spermatogonial cells, gastrointestinal system, pancreas, kidney, liver, adrenal and pituitary glands, bone tissue, cardiovascular system, immune cells [4].

STRUCTURE OF ERA

Nuclear estrogen receptors consist of two subtypes- ERα and ERβ. Both ERα and ERβ proteins can be activated by 17β-estradiol (E2), and they share similar chemical structures. The structure of ERα is composed of five domains including A/B, C, D, E, F-domain. The A/B domain (NH2-terminal) is the site of the transcriptional activation which has also been called as activation function 1 (AF1, the coactivator). The C-domain is the DNA-binding site and regulate the target gene transcription though estrogen response element on the DNA, which is known as AGGTCAnnnTGACCT. The three remaining domains (D/E/F domain) constitute ligand-binding domain, and contain a nuclear localization signal which exists in D-domain or hinge region. The E-domain is also another site with transcriptional activation function (i.e., AF-2, the coactivator) and can interact with a few coregulators. The role of carboxyl-terminal F-domain may be involved in a complex regulation of its transcriptional activities [5].

Although the highly homologous structure in the ligand binding domain between the two nERs lead to their similar affinities to E2, there are variable binding affinities to other ligands. ERα is easier to bind with estrone, and ERβ tends to bind with estriol.

FUNCTION AND CLINICAL ASPECTS OF ESTROGEN-ERA IN ENDOCRINE SYSTEM

The classical actions of estrogens are known as their positive and/or negative feedback regulation of hypothalamus-pituitary-gonad axis (i.e., ovary) activities and the stimulation of prolactin secretion, which are mainly mediated by ERα. However, the influences of estrogens on other endocrine tissues have received more and more attentions. For example, although surviving and reproduction are the most important goals for all livings, metabolic processes, which are mainly regulated by endocrine system, are the basis for all life activities. The metabolism is regulated to keep a series of balance between catabolism and anabolic, oxidation and reduction etc.

The modulation of metabolic processes by estrogens and the ER subtypes have been investigated by a few studies.

In this chapter, we would focus on the influences of estrogens on other endocrine tissues than hypothalamus-pituitary-gonad axis, especially those mediated by ERα.

The Potential Impact of Estrogen- ERα on Thyroid

The Effect on Thyroid Function

The hypothalamic-pituitary-thyroid (HPT) axis is important in cell metabolism. Thyroxine (T4) and triiodothyronine (T3), produced by thyrocytes, have been known to increase the basal metabolic rate. The basal metabolic rate of men is higher than that of women [6] and the former seem to be susceptible to oxidative stress [7-10]. The indirect impact of estrogens on thyroid function may be related to the elevated serum level of thyroxine binding globulin (TBG), which is induced by estrogens and can lead to high blood concentration of total T4 (TT4) and total T3 (TT3) [11]. Hypothyroid women receiving estrogen therapy need more T4 [12], which may be due to estrogen-induced more TBG in the blood. ERα expression has been found in the hepatic cells of healthy subjects and patients with liver diseases, while ERβ expression mainly in the latter [13-14]. It suggests that ERα may mediate the upregulation of TBG production by estrogens.

In addition, Redox plays an important role in the synthesis of thyroid hormones, which is involved in the production of hydrogen peroxide (H2O2) and the regeneration of thyroperoxidase (TPO) activity. TPO is the key enzyme in bio synthesis of thyroid hormone [15], which consumes H2O2. The dual oxidase (DUOX) is localized at the apical membrane of thyroid follicular cells and is the main enzyme responsible for the production of H2O2 in thyroid tissue. It has been found that E2 can stimulate the expression of DUOX and the production of H2O2 through the activation of ERα [16-17]. Especially, in one study with female rats, exogenous E2 significantly elevated the activity of TPO [18].

The above findings suggest ERα activation may affect serum levels of thyroid hormones, especially under high estrogen conditions. It should be taken into account in those females with hyperthyroxinemia whether they get pregnant or have been taking the medicine containing estrogens.

The Effect on Autoimmune Thyroid Disease

Autoimmune thyroid disease (AITD) is characterized by high expression of antithyroid autoantibodies and local lymphocyte infiltration. It mainly comprises of Hashimoto thyroiditis and Graves' disease. Numerous immunocytes, including helper T (Th) cells, regulatory T (Treg) cells, autoreactive B cells, natural killer (NK) cells and dendritic/ macrophages, have been involved in the pathogenesis of AITD. Through the ERs, estrogens can regulate both immunocytes of innate immune system (neutrophils, monocytes /macrophages, etc.) and those of adaptive immune system (T cells and B cells) [19-23]. T cells, including $CD4^+$ and $CD8^+$ cells, can be modulated by estrogen [24-27]. It has been known that high levels of estrogens (e.g., during pregnancy) are able to alter the immune response from Th1 to Th2 type [28-30]. The frequency and expansion of Treg cells can also be promoted by estrogens, which play a crucial role in the down-regulation of immune responses [31-32]. Estrogens are able to influence the function, differentiation, activity of B cells [33-34].

It has been shown that estrogens increase numbers of plasma cell and autoantibody-producing cells [33, 35]. Almost all immunocyte, which may promote the development of AITD, are regulated by estrogens and their receptors [36]. Much indirect evidence has also suggested that estrogens have important effects on AITD [36]. For example, a cross-sectional population-based study with a total of 4109 adult males, higher ratios of E2 to T were vitally related to development of AITD in males [37]. In a recent study, we have directly demonstrated that low physiological levels of E2 can stimulate the occurrence and development of autoimmune thyroiditis. Most of the immunocytes express ERα. Especially there is more ERα expression than ERβ in $CD4^+$ T cells, and both the two ER subtypes are low in $CD8^+$ while more ERβ expression than ERα in B cells [23]. It has been considered that the ERα activation usually exhibits more prominent effects than ERβ.

Activated ER α can promote the progress of those autoimmune inflammatory responses which are mainly mediated by T cells [38]. It has been found that ERα activation plays the predominant and immunostimulatory role in estrogen-mediated modulation of lupus while ERβ activation appears to have a slightly immunosuppressive effect on this disease. ERα activation coincidentally increased serum prolactin concentrations, which may also contribute to the acceleration of the disease activity (Figure 1) [39].

However, the immunoregulatory actions of ERα and ERβ may be disease-specific. ERα knockout mice have shown immune complex-mediated glomerulonephritis, indicating the potential immnuo-stimulatory activity of ERβ [23].

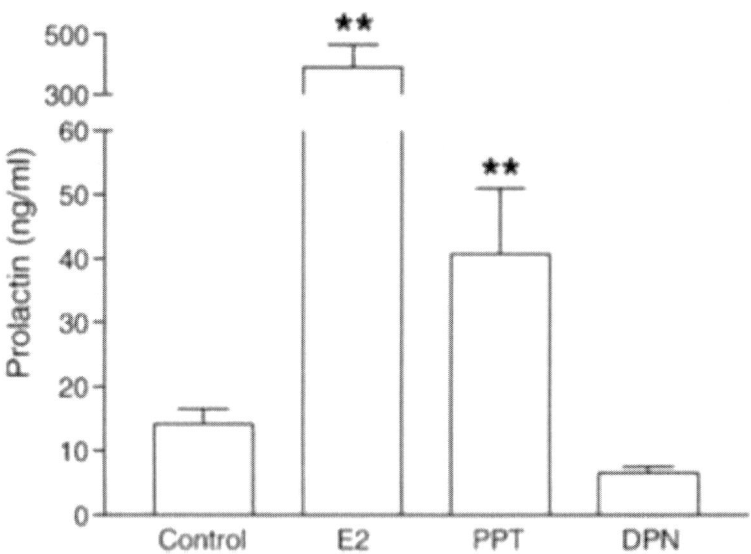

Figure 1. Serum concentration of prolactin in ovariectomized B/W mice treated with E2 (ER subtype-nonselective agonist), PPT (ERα-selective agonist), DPN (ERβ-selective agonist) or control vehicle. Prolactin in the sera from those representative 18-weekold B/W mice (n = 6–9/group) was measured by RIA. Data are shown as mean ±S.E.M. $^{**}P < 0.01$ vs. vehicle control group [39].

Qin et al. have recently reported that ERβ may be the predominant ER subtype responsible for estrogen-mediated development of autoimmune thyroiditis, and its activation may enhance Th17-type responses through genomic pathways and alteration of APCs' activities [40]. ERα expression has been found in the retro-orbital fibroblasts from Graves' ophthalmopathy patients, which may mediate the immunomodulation by estrogens [41]. The in vitro experiment has suggested that glucocorticoids can inhibit the gene expression of ERα in those orbital fibroblasts with dose-related dependence [41], which may represent one mechanism for glucocorticoids to treat Graves' ophthalmopathy [41].

The Effect on Thyroid Cancer

Thyroid carcinoma (TC) is classified into four pathological types, including papillary thyroid cancer (PTC), follicular thyroid cancer (FTC), medullary thyroid cancer (MTC), and anaplastic thyroid cancer (ATC). PTC, one type of the differentiated thyroid carcinoma (DTC), accounts for majority of the TC cases. The incidence rate of TC has remarkably grown in recent years [42]. There is no obvious gender difference in the incidence rates of TC (IRTC) before puberty. However, with the increasing production of estrogen after puberty, the IRTC goes up, and presents gender difference. After menopause, the IRTC of females obviously falls back and is equivalent to that of males. The evidence above indicates that estrogen promote the increase of IRTC [43]. Estrogen is considered as a powerful growth factor for thyroid tumor (benign or malignant) [44]. A variety of studies on ER expression have been performed in both normal and neoplastic thyroid tissues with extremely heterogeneous results. By immunohistochemistry (IHC) staining of clinical samples, several studies reported that the expression of ERα were obviously higher in patients with thyroid neoplasms (TN) than that in normal controls [36]. The expression of ERβ was obviously lower in various types of TN tissues than in the normal gland, especially in poorly DC and undifferentiated carcinoma of thyroid [45]. We have found that ERα expression was significantly up-regulated in female PTC patients as compared with that of nontoxic goiter patients (Figure 2) [46].

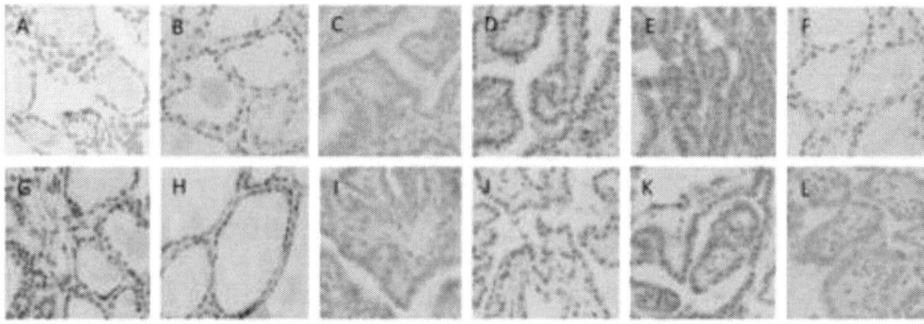

Figure 2. Immunohistochemical staining of ERα and ERβ1 in PTC lesions and NTG tissue (400×). ERα and ERβ1 expression was analyzed by immunohistochemical staining of formalin-fixed and paraffin-embedded thyroid tissue sections from patients with papillary thyroid cancer (PTC) and those with nodular thyroid goiter (NTG) using specific antibodies and the Elivision™ plus two-step system. (A) NTG tissue with negative ERα staining; (B) NTG tissue with positive nuclear ERα staining; (C) PTC tissue with negative ERα staining; (D) PTC tissue with positive nuclear ERα staining; (E) PTC tissue with positive nuclear and cytoplasmic ERα staining; (F) NTG tissue with negative ERβ1 staining; (G) NTG tissue with positive nuclear ERβ1 staining; (H) NTG tissue with positive nuclear and cytoplasmic ERβ1 staining; (I) PTC tissue with negative ERβ1 staining; (J) PTC tissue with positive nuclear ERβ1 staining; (K) PTC tissue withpositive nuclear and cytoplasmic ERβ1 staining; (L) PTC tissue with positive cytoplasmic ERβ1 staining [46].

However, another study observed an association between ERα expression and pathological characteristics of thyroid, and got an opposite result that the expression of ERα were related to the remission of PTC [47]. To further investigate the correlations between ERα expression and the clinicopathological characteristics of TN, we have performed a meta-analysis with the results that ERα expression was significantly higher in thyroid cancer than in non-cancer neoplasms, but ERβ expression did not show difference (manuscript in preparation). The activation of ERα can impact the tumors' biological behaviors, and may alter the prognosis of TC patients. ERα activation was able to induce the proliferation of PTC [48]. Zeng et al. have found that ERα activation can stimulate the proliferation of TN cells by ERK1/2-related signaling pathway [49]. They further found that the regulatory actions mediated by ERα and ERβ were opposite, and ERαactivation promoted while ERβ suppressed the proliferation of TN cells [48] in addition, our study has shown that ER α activation enhanced while ERβ activation suppressed the migration and invasion activities of a typical

PTC cell line [50]. Some clinical also studies suggest that ERα expression is an important prognostic marker for lymph node metastasis and more invasive phenotypes in TN [51-52]. Thus, ERα mediates the stimulatory actions of TN (especially PTC) development by estrogens. It can become a new target in the diagnosis, prognosis and treatment of PTC.

Pancreatic Islet and Glucose Metabolism

Contemplating why metabolic processes differ by gender will facilitate us to understand the molecular function of estrogen and ER. Males preferentially accumulate visceral fat mass [53-54], while females tend to accumulate more fat in the subcutaneous adipose tissue (SAT) prior to menopause for maintaining reproductive capacity. E2 plays a prominent role in the distribution of adipose tissue in a SAT pattern, which protect female from risks of multiple vascular diseases [55]. White adipose tissues (WAT), storing triacylglycerol, is the major energy reserve during imbalance of food intake and exercise, while brown adipose tissues (BAT) dissipate chemical energy in the form of thermogenic respiration. Pre-menopausal women have more BAT and are more sensitive to BAT activation than men or postmenopausal women. One study suggests that the browning of adipocytes can be induced by ERα [56]. ERα activation can regulate the amount of WAT in both males and females [57]. Increase of adiposity has been observed in ovariectomized animals [58]. Vascular endothelial growth factor A (VEGF-A) is a critical factor in the regulating of angiogenesis in adipose tissue. Poor vascularization in adipose tissue is related to insulin resistance (IR). ERα mediates the regulation of both adipose tissue function and VEGFA expression in adipose tissue by E2 [59]. Estrogens are critical in the control of glucose homeostasis [55, 60-61]. Estrogens have been found to improve IR and pancreatic β cell function [55, 62], in which ERα and ERβ may play different roles. ERα activation can prevent the damage of pancreatic beta cell due to oxidative stress, amyloid polypeptide toxicity, gluco-lipotoxicity and apoptosis, and induce their proliferation [59, 63-66]. Using ERα knockout mice, it has been demonstrated that estrogens have important regulatory actions on the insulin sensitivity in the whole body through ERα [67]. It has also been found that ERβ activation can increase both the insulin secretion from β islets after glucose stimulation and pancreatic β-cell mass, and further improve blood glucose and insulin sensitivity through restoring first-phase insulin secretion [68]. Both in vitro and in vivo studies showed that hyperglycemia, at least moderate to severe, was correlated with increased expression of ERα mRNA. The absence of ERα in β islet predisposed oxidative stress-induced diabetes in mice [69]. However, ERα activation may promote the hypertrophy of diabetic glomeruli by inducing the production of transforming growth factor β [70]. Generally, ERα may mediate protective actions on the development of type 2 diabetes, and lack of estrogens may promote IR and hyperglycemia.

Others

In addition to thyroid and pancreatic islet, ERs are expressed in several endocrine tissues, such as hypothalamus, pituitary, adrenal cortex [71-72]. ERα in pituitary is related to pulsatility of luteinizing hormone (LH) at mid-gestation in kivizcacha [73]. E2 mediate the regulation of follicle-stimulating hormone (FSH) and LH by both ERβ and GPER1 in the pituitary [74]. ERα also regulate the progression of prolactinoma [75-76], since its activation can stimulate the secretion of prolactin (Figure1) [39]. Both ERβ and ERα are expressed in the hypothalamus with a relatively high level of ERβ [71]. They may mediate the regulation of estrogens on body temperature, and food intake in the hypothalamus [77]. Sex difference also exists in the affection of appetite. In an animal study, orchiectomy decreased food intake by reducing meal frequency, and ovariectomy enhanced food intake by increasing meal size [78-79]. One important mechanism may be involved in Ghrelin. ERα and ghrelin are co-localized in the hypothalamus, such as arcuate nucleus [80-81]. Ghrelin can directly act on ERα-expressing neurons in ARC nucleus [81]. The activation of ERα can inhibit the activities of AgRP\NPY neurons, and inactivation of ERα may enhance the appetite and lead to the increase of body weight [82-84]. It has been recognized that estrogens can stimulate the hypothalamus-pituitary-adrenal (HPA) axis if no proinflammatory stimulus is received, while it suppresses the activity of HPA axis in the individuals under the conditions of proinflammatory stimuli [85]. However, there is lack of study on the responsible ER subtypes in this field. The expressions of ERα and ER-β are found in all the adrenal cortex zones of the fetus, and they mediate the effects of estrogens on fetal adrenal function [80]. Estrogens may either stimulate or suppress the responsivity of adrenal cortex to ACTH, and especially affect the production of corticosteroids, which depends on circulatory level of estrogens [86] Under the conditions of low estrogen, the adrenal cortical site in the fetus was significantly increased [86]. These findings suggest the potential effects of xenoestrogen on the development of human adrenal gland, which are widely existent in our environments.

Estrogens have been known to regulate the synthesis of lipoproteins. The inactivation of hepatic ERα may lead to the accumalatio of triglycerides in the liver [87]. Circulatory estrogens may affect the activity of hepatic ERα, and regulate the transport of cholesterol. Hepatic ERα inactivation may account for the susceptibility of dyslipidemia of postmenopausal women [88].

Altogether, estrogens have multiple effects on the functions and growth of endocrine tissues through the activation of ERα and ERβ besides the classical influences on hypothalamus-pituitary- gonad axis. And different ER subtypes may mediate different actions of estrogens. ER subtype-specific agonist or antagonist may become new strategies for the treatment of estrogen-dependent endocrine and metabolic disorders.

REFERENCES

[1] Menasce, L. P. et al. *Genomics,* 1993; 17, 263 - 265.
[2] Mosselman, S. et al. *FEBS Lett.,* 1996; 392, 49-53.
[3] Enmark, E. et al. *J. Clin. Endocrinol. Metab.,* 1997; 82, 4258 - 4265.
[4] Vrtacnik, P. et al. *Biochem. Med.*, (Zagreb). 2014; 24, 329 - 342.
[5] Nilsson, S. et al. *Physiol. Rev.,* 2001; 81, 1535 - 1565.

[6] Ferraro, R. et al. *J. Clin. Invest.,* 1992; 90, 780 - 784.

[7] Barp, J. et al. *Braz. J. Med. Biol. Res.,* 2002; 35, 1075 - 1081.

[8] Ide, T. et al. *Arterioscler. Thromb Vasc. Biol.,* 2002; 22, 438 - 442.

[9] Bhatia, K. et al. *Am. J. Physiol. Regul. Integr. Comp. Physiol.*, 2012; 302, R274 - 282.

[10] Matarrese, P. et al. *Antioxid. Redox. Signal,* 2011; 15, 2825 - 2836.

[11] Ain, K. B. et al. *J. Clin. Endocrinol. Metab.,* 1987; 65, 689 - 696.

[12] Arafah, B. M. *N Engl. J. Med.,* 2001; 344, 1743 - 1749.

[13] Rossini, G. P. et al. *Gastroenterology,* 1989; 96, 1102 - 1109.

[14] Iavarone, M. et al. *Cancer,* 2003; 98, 529 - 534.

[15] Fortunato, R. S. et al. *J. Clin. Endocrinol. Metab.,* 2010; 95, 5403 - 5411.

[16] Fortunato, R. S. et al. *J. Endocrinol.,* 2010; 221, R31 - 40.

[17] Stepniak, J. et al. *Int. J. Mol. Sci.,* 2018; 19.

[18] Lima, L. P. et al. *Steroids,* 2006; 71, 653 - 659.

[19] Kovats, S. *Cell Immunol.,* 2015; 294, 63 - 69.

[20] Kovats, S. *Horm. Behav.,* 2012; 62, 254 - 262.

[21] Polanczyk, M. J. et al. *J. Immunol.,* 2004; 173, 2227 - 2230.

[22] Grimaldi, C. M. et al. *J. Clin. Invest.,* 2002; 109, 1625 - 1633.

[23] Khan, D. et al. *Front. Immunol.,* 2015; 6, 635.

[24] Lelu, K. et al. *J. Immunol.,* 2011; 187, 2386 - 2393.

[25] Priyanka, H. P. et al. *Mol. Immunol.,* 2013; 56, 328 - 339.

[26] Robinson, D. P. et al. *J. Virol.,* 2014; 88, 4711 - 4720.

[27] Karpuzoglu-Sahin, E. et al. *Cytokine,* 2001; 14, 208-217.

[28] Sabahi, F. et al. *Am. J. Reprod. Immunol.,* 1995; 33, 381 - 393.

[29] Marzi, M. et al. *Clin. Exp. Immunol.,* 1996, 106, 127 - 133.

[30] Matalka, K. Z. *Neuro Endocrinol. Lett.,* 2003; 24, 185 - 191.

[31] Polanczyk, M. J. et al. *J. Neuroimmunol.,* 2005; 170, 85 - 92.

[32] Tai, P. et al. *J. Cell Physiol.*, 2008, 214, 456 - 464.

[33] Verthelyi, D. I. et al. *Cell Immunol.,* 1998; 189, 125 - 134.

[34] Grimaldi, C. M. et al. *J. Immunol.,* 2006; 176, 2703 - 2710.

[35] Bernardi, A. I. et al. *Immun. Inflamm. Dis.,* 2014; 2, 214 - 225.

[36] Lu, Y. et al. *Minerva Med.,* 2016; 107, 239 - 244.

[37] Chen, Y. et al. *Thyroid,* 2017; 27, 960 - 966.

[38] Mohammad, I. et al. *Sci. Signal,* 2018; 11.
[39] Li, J. et al. *Clin. Immunol.*, 2007; 123, 219 - 226.
[40] Qin, J. et al. *Clin. Immunol.*, 2018; 190, 41 - 52.
[41] Cury, S. S. et al. *Arch. Endocrinol. Metab.,* 2015; 59, 273 - 276.
[42] Wiltshire, J. J. et al. *Thyroid,* 2016; 26, 1541 - 1552.
[43] Malespin, M. et al. *Clin. Liver Dis.,* 2019; 23, 233 - 246.
[44] Derwahl, M. et al. *Endocr. Relat. Cancer*, 2014; 21, T273 - 283.
[45] Liu, J. et al. *Pathol. Res. Pract.,* 2014; 210, 830 - 835.
[46] Huang, Y. et al. *BMC Cancer*, 2014; 14, 383.
[47] Sturniolo, G. et al. *Eur. Thyroid J.,* 2016; 5, 224 - 230.
[48] Chen, G. G. et al. *Curr. Cancer Drug Targets,* 2008; 8, 367 - 377.
[49] Zeng, Q. et al. *Cell Prolif.,* 2007; 40, 921 - 935.
[50] Dong, W. et al. *Int. J. Endocrinol.,* 2013; 2013, 941568.
[51] Ahmed, R. A. et al. *Pathol. Oncol. Res.,* 2015; 21, 793 - 802.
[52] Magri, F. et al. *Endocrine,* 2015; 49, 429 - 435.
[53] Palmer, B. F. et al. *Mol. Cell Endocrinol.,* 2015; 402, 113 - 119.
[54] Bracht, J. R. et al. *Ann. N Y Acad. Sci.,* 2019; 1461, 127 - 143.
[55] Hamilton, D. J. et al. *Physiol. Rep.,* 2016; 4.
[56] Santos, R. S. et al. *Mol. Metab.,* 2018; 18, 51 - 59.
[57] Heine, P. A. et al. *Proc. Natl. Acad. Sci. U S A*, 2000; 97, 12729 - 12734.
[58] Wallen, W. J. et al. *J. Nutr.,* 2001; 131, 2351 - 2357.
[59] Fatima, L. A. et al. *Sci. Rep.,* 2017; 7, 16716.
[60] Vasudevan, N. et al. *Physiol. Rev.,* 2002; 82, 923 - 944.
[61] Gupte, A. A. et al. *J. Diabetes Res.,* 2015; 2015, 916585.
[62] Riant, E. et al. *Endocrinology,* 2009; 150, 2109 - 2117.
[63] Shaklai, S. et al. *Horm. Metab. Res.,* 2018; 50, 485 - 490.
[64] Ren, Z. et al. *Cell Biol. Int.,* 2010; 34, 523 - 530.
[65] Zhou, Z. et al. *J. Biol. Chem.,* 2018; 293, 4735 - 4751.
[66] Tiano, J. P. et al. *J. Clin. Invest.,* 2011; 121, 3331 - 3342.
[67] Manrique, C. et al. *Cardiorenal Med.,* 2012; 2, 200 - 210.
[68] Alonso-Magdalena, P. et al. *Diabetes,* 2013; 62, 2015 - 2025.
[69] Mukai, M. et al. *J. Pediatr. Surg.,* 2012; 47, e1 - 4.

[70] Lovegrove, A. S. et al. *Am. J. Physiol. Renal Physiol.,* 2004; 287, F586 - 591.
[71] Register, T. C. et al. *Brain research,* 1998; 788, 320 - 322.
[72] Kaludjerovic, J. et al. *J. Nutr. Metab.,* 2012; 2012, 837901.
[73] Proietto, S. et al. *Gen. Comp. Endocrinol.,* 2019; 273, 40 - 51.
[74] Qin, X. et al. *Front. Endocrinol.,* (Lausanne). 2018; 9, 308.
[75] Liu, J. et al. *Biosci. Rep.,* 2020; 40.
[76] Wang, C. et al. *Mol. Med. Rep.,* 2018; 18, 4037 - 4041.
[77] Kelly, M. J. et al. *Vitam. Horm.,* 2005; 71, 123 - 145.
[78] Asarian, L. et al. *Philos. Trans. R Soc. Lond. B Biol. Sci.,* 2006; 361, 1251 - 1263.
[79] Hirschberg, A. L. *Maturitas,* 2012; 71, 248 - 256.
[80] Frazao, R. et al. *Am. J. Physiol. Endocrinol. Metab.,* 2014; 306, E606 - 614.
[81] Shen, X. et al. *Acta Veterinaria et Zootechnica Sinica,* 2017; 48, 2399 - 2406.
[82] Olofsson, L. et al. *Proc. Natl. Acad. Sci. U S A,* 2009; 106, 15932 - 15937.
[83] Xu, P. et al. *J. Clin. Inuest.,* 2015; 125, 2861 - 2876.
[84] Eckel, L. A. *Physiol. Behav.,* 2011; 104, 517 - 524.
[85] Straub, R. H. *Endocr. Rev.,* 2007; 28, 521 - 574.
[86] Albrecht, E. D. et al. *Endocrinology,* 1999; 140, 5953 - 5961.
[87] Della Torre, S. et al. *Cell Rep.,* 2016; 15, 360 - 371.
[88] Gerhardt, K. *Biol. Reprod.,* 2016; 94, 131.

In: Estrogen Receptors
Editors: George G Chen et al.
ISBN: 978-1-53618-228-6

Chapter 5

ESTROGEN SIGNALING PATHWAY IN LUNG CANCER AND ITS IMPACT ON THERAPY

***Vianey Rodriguez-Lara*[*], PhD**
Tissue and Cellular Biology Department, School of Medicine,
Universidad Nacional Autonoma de México (UNAM),
Mexico City, Mexico

ABSTRACT

Lung cancer is the leading cause of cancer related-deaths worldwide. Nowadays, lung adenocarcinoma (ADC) is the most common type of lung cancers. Recently, it has been recognized that lung ADC is a hormone-dependent cancer, and estrogens play an important role in lung carcinogenesis and tumor progression. The high expression of aromatase enzyme as well as estrogen receptor beta has been identified in lung ADC tissues from men and female patients. The expression of both proteins has been associated with worse prognosis and poor survival in patients with lung ADC. Estrogen promotes lung carcinogenesis by activating several signaling pathways and promoting the expression of genes mainly related to growth factors and involved with cell proliferation and cell cycle. Although some mechanisms of estrogens in lung carcinogenesis have been identified, the role of these hormones in lung cancer continues to be

[*] Corresponding Author's Email: vianeyrodriguezlara@yahoo.com.mx.

investigated. Because of the relevance of the estrogen in lung cancer, the use of antiestrogenic therapy has been considered in lung cancer treatment. There are currently ten clinical trials in process to test the efficacy of this kind of the therapy in lung cancer, hoping that anti-hormonal therapy will be applied in lung cancer treatment in the future.

Keywords: lung cancer, women, estrogen, anti-hormonal therapy

INTRODUCTION

Globally, lung cancer is an important health problem due to the fact that it exhibits the highest incidence and mortality of all types of cancers. Among males, it continues to be the leading cause of cancer related deaths; while in women lung cancer has significantly increased in the last decades and nowadays it has the second place in mortality [1-3]. Currently, the most common type of lung cancer worldwide is the non-small cell lung cancer (NSCLC) and the lung adenocarcinoma (ADC) is the predominant subtype of NSCLC [4].

Despite the advances in treatment and early diagnosis, the survival of patients with NSCLC remains poor; 80% of tumors are detected in advanced stages and have poor prognosis. Currently the 5-year overall survival of patients with lung cancer is less than 20% [5].

Several differences in lung cancer presentation, histology, behavior and response to treatment have been observed by sex. For instance, a high percent of women that present lung cancer are non-smokers (53%) compared whit the low percent of men non-smokers who develop this neoplasm (15%) [6]. Also, to the same intensity of smoking habit, women exhibit lung cancer earlier than men [7]. In addition, *EGFR* and *ALK* mutations are more common in women than in men and women have the highest response to target therapy based in tyrosine kinase inhibitors (TKI) and ALK inhibitors [8-10] and chemotherapy response is also higher in women compared to men [9], nevertheless men respond better to immunotherapy [11].

Moreover, when hormonal status has been considered, premenopausal women exhibit mainly lung adenocarcinoma tumor type, whom are diagnosed in advanced stages, with distant metastases, with poorly differenced tumors and worse prognoses compared to postmenopausal women and men [12-14]. Recently our group has reported that premenopausal women with NSCLC exhibit significantly lower overall survival compared to postmenopausal women and men [15].

All these data together support that NSCLC mainly ADC is a different pathology in women and men and it is strongly influenced by hormones. Currently lung ADC is considered as a hormone dependent cancer, and estrogen promote lung carcinogenesis and tumor progression, impacting significantly the prognosis of patients.

ROLE OF ESTROGEN IN LUNG CANCER

Estrogens are steroid sexual hormones that are produced mainly by the aromatase enzyme in ovaries, the placenta during pregnancy and in lower concentration are also produced by adrenal glands. In addition to their sexual function, estrogens regulate several functions in non-sexual organs such as lungs. Estrogens through their alpha (ERα) and beta (ERβ) receptors allow the complete formation of alveolar units during embryonic development and can induce alveolar regeneration in adult mice. The biological function of estrogen through each receptor is different; while the response through ERα is essential for the formation of the alveolar surface, through ERβ the development of the extracellular matrix is modulated promoting the elasticity of lung tissue. In addition, the signaling pathway through ERα but not ERβ stimulates the inflammatory response [16-18]. Estrogens support lung development and its physiology but also have a role in pathology.

Several studies have shown the importance of estrogen in lung carcinogenesis, and in prognosis of lung cancer patients and currently estrogens have had an important impact on lung cancer therapy.

ESTROGENS IN NSCLC

Lung cancer cells are responsive to circulating estrogen, however they are also able to produce estrogen through the aromatase enzyme. Aromatase overexpression has been detected in lung cancer cell lines and in 86% of lung tumors from both women and men [19, 20]. A higher aromatase expression has been reported in tumor area compared to health lung tissue. Additionally, a higher concentration of estrogen has been observed in tumor tissue [21], supporting that intratumoral estrogen levels are produced locally by aromatase enzyme. Moreover, metastatic tumor exhibits higher estrogen concentration compared to the detected in primary tumor, suggesting that local production of estrogen can support metastases in lung cancer [19, 21].

Aromatase overexpression and the high estrogen levels have been related to patient prognosis. Premenopausal women with NSCLC who had high estrogen levels, exhibited poor differentiated tumors, advanced stages and poor survival compared to postmenopausal women and men [14, 15, 22]. Also, men with NSCLC that had higher levels of estrogen showed worse prognosis compared to men who exhibited lower levels of estrogen [23]. These data show that regardless the sex of the patient, tumor aromatase expression and intratumoral estrogen levels can stimulate the lung carcinogenesis.

The effect of exogenous estrogen intake in lung cancer has been investigated, however results are still contradictory. For example, some studies have shown that the use of replace hormonal therapy (HRT) based on estrogen, decreases the risk of lung cancer; while other studies have reported a protective effect or no effect of hormonal therapy use [24-26]. The adverse effects of estrogen intake in prognosis of lung cancer patients have been reported. In this regard, several studies have concluded that HRT based on estrogen, increases the incidence of lung cancer [27-29]. A prospective study that involved 36,588 women that used combined estrogen plus progestin HRT showed 50% increase risk to develop lung cancer after 10 years of HRT use [28]. In addition, the women health initiative (WHI) investigated the effect of combined HRT and showed no increased incidence of lung cancer, but an increased mortality due to lung cancer was significantly noted. Also, it was observed poorly differentiated tumors and distant metastases in women who used combined HRT [30]. In addition, lower survival was observed in women who use HRT compared to women who did not take the HRT (39 *vs* 79 months) [31]. The risk of lung cancer decreased after hormonal treatment cessation [32]. Although the effect of HRT is still in contradiction, several studies have showed the adverse effect of estrogen in lung cancer.

ESTROGEN RECEPTOR IN NSCLC

NSCLC cells are sensitive to estrogen due to the expression of the estrogen receptor; both nuclear form of the receptor; estrogen receptor alpha (ERα) and estrogen receptor beta (ERβ) have been identified in NSCLC cell lines and in tumors from patients. ERβ is the most important form of estrogen receptor in lung cancer, since it is overexpressed in 60 to 80% of tumors from male and female patients [14, 33-35]. ERβ and aromatase tumor expression has been related to worse survival rate of NSCLC patients [36]. Additionally, ERβ expression has been related to mutations in 500 genes [37]. In contrast, ERα is expressed only in approximately 23% of patients with lung cancer [38], these patients showed substantial survival benefit. Also, ERα expression has been related only to 20 mutated genes [37].

Although ERβ is expressed in tumors from women and men, its expression has been associated with hormonal status, since higher expression was observed in younger premenopausal women compared to postmenopausal women and men [14].

Five isoforms of ERβ have been identified, however only the ERβ1 isoform is able to bind the ligand and as a homodimer activates the estrogen dependent pathways, other ERβ isoforms are inactive, but by forming heterodimers with ERβ1 can regulate transcriptional activity [39]. ERβ1 expression has been identified in tumors from women and men and has been related to aggressive tumor behavior [40].

On the other hand, the G protein coupled receptor for estrogen (GPR30) has also been identified in NSCLC, mainly in the cytoplasm of tumor cells, which show higher concentration of GPR30 mRNA and higher activity of receptor compared to healthy lung cells [41]. Estrogen promotes tumor cell proliferation, invasion and migration through GPR30 receptor, while the treatment with G15 an antagonist of GPR30, suppressed this effect. Also GPR30 expression has been associated to IIIA-IV lung cancer stages, metastases to lymph nodes and poorly differentiated tumors [42, 43].

Estrogen Mechanisms in Lung Carcinogenesis

Estrogen promotes lung carcinogenesis and tumor progression through several pathways (Figure 1). Estrogen metabolites generate reactive oxygen species (ROS) that cause oxidative DNA damage and form DNA adducts producing mutations associated with NSCLC [44, 45]. In addition, estrogen stimulates the transcription of estrogen dependent genes in genomic pathway through its binding to the nuclear estrogen receptor. In NSCLC, estradiol promotes cell proliferation and cell cycle progression inducing *c-myc* and *D-cyclin* gene expression [45, 46]. Through non-genomic pathway, estrogen activates kinases proteins, ionic channels and second messengers. In NSCLC, estrogen activates AMPc, PI3K, MAPK, AKT, ERK and EGFR pathways promoting cell proliferation and tumor growth [47, 48].

Epidermal growth factor receptor (EGFR) pathway has a significant relevance in lung cancer, since up to 89% of patients with NSCLC overexpress this receptor and it is one of the mainly mutated genes in NSCLC. Activation of EGFR increases cell survival, cell proliferation and cell migration, and it also promotes angiogenesis and metastases [49]. A functional relationship between ER and EGFR pathways has been identified in NSCLC, since estrogen stimulation activates EGFR pathway independently of EGF availability, suggesting a non-nuclear transactivation of EGFR. Also, EGFR activation, promotes aromatase expression in lung cancer cells [47], in consequence, the increases of estrogen production and estrogen receptor activation was observed. Moreover, estrogen pathway stimulates EGF production and EGFR activation [50, 51], establishing a positive feedback between both signaling pathways. Bidirectional signaling between ER and EGFR could have an important therapeutic advantage due to the antiestrogenic treatment results in a blockage of both EGFR and ER pathways [52, 53].

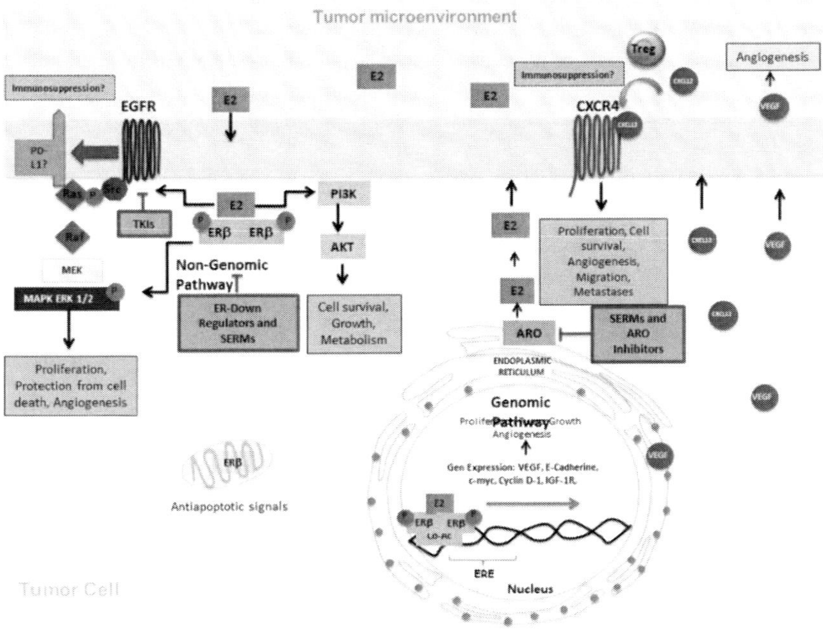

Figure 1. Estrogen pathway signaling through ERβ in NSCLC and its potential as a target for lung cancer therapy. Estrogen stimulates lung carcinogenesis through genomic and non-genomic pathways; the genomic pathway activation results in estrogen dependent genes expression that promotes proliferation, tumor growth and angiogenesis; while through the non-genomic pathway, estrogen promotes proliferation, angiogenesis and cell survival through the activation of EGFR, MEK, ERK, PI3K, AKT signaling pathways. Also estrogen could establish an immunossuppresive microenvironment through CXCR4/CXCL12 and PD-L1 pathways, stimulating tumor immune evasion and progression. The gray boxes show the cellular events stimulated by estrogen in NSCLC. Antiestrogen therapy could decrease the estrogen effect on lung cancer cells that express estrogen receptor. Red boxes show the targets of antiestrogen therapy and the events that could be blocked. Combined treatment based in antiestrogen therapy and tyrosine kinase inhibitor (TKI) has improved the response to treatment due to the relationship between these pathways. Estrogen could stimulate immune evasion through PD-L1 and CXCR4/CXCL12 pathways, thus antiestrogen therapy could improve the response to immunotherapy.

It is interesting that EGFR mutation has been related to ERβ expression in NSCLC, Nose and coworkers (2009) showed that 67% of EGFR mutation positive tumors exhibited high ERβ nuclear expression compared to 37% in EGFR wild-type tumors [52].

Other mechanism of estrogen in lung carcinogenesis is related to tumor cell survival. ERβ has been identified in the mitochondria of NSCLC and has been associated to anti-apoptotic signals. ERβ inhibits Bad function by disrupting Bad-Bcl-X (L) and Bad-Bcl-2 interactions, as well as C-cytochrome release, protecting tumor cells from apoptosis [54, 55].

Estrogen promotes angiogenesis in NSCLC by stimulating VEGF expression and activating VEGFR. A relation between VEGFR and EGFR pathways has been observed due to activation of EGFR results in VEGF production by lung cancer cells and endothelial cell proliferation [45]. The combined fulvestrant (antiestrogen drug) and vandetanib (anti-angiogenic drug) treatment decrease the cell proliferation and angiogenic activity supporting the relationship between ER and VEGFR pathways in lung cancer [56].

ER pathway has also been related with fibroblast growth factor receptor (FGFR) NSCLC. Mice exposed to tobacco carcinogens and treated with the antiestrogen drugs fulvestrant and anastrozol, exhibited a decreased production of fibroblast growth factor (FGF) -2 and -9 as well as a reduced expression of SOX2 and Nanog, suggesting that estrogens might stimulate FGF-2 and 9 secretion and promote the appearance of cancer stem cells markers in lung cancer cells [57]. Additionally, estrogen promote insulin like growth factor-1 (IGF-1) secretion, promoting the proliferation of A549 lung cancer cells [58] as well as the expression of E-cadherin [59].

Migration and metastases is also stimulated by estrogen through osteopontin expression. The relationship between ERβ, EGFR and osteopontin has been identified due to the treatment with EGFR and ER antagonists inhibited the osteopontin dependent cell migration [12].

Estrogen modulates tumor immune response by modifying the microenvironment. Estrogen stimulates pro-inflammatory cytokines secretion, attracts cells that promote tumor growth and drives tumor cell migration and metastases [60]. In NSCLC, macrophages associated to the tumor (TAMs), express aromatase and in response to estrogen, produce VEGF that stimulate the recruitment of immune cells which produce anti-inflammatory cytokines promoting an immunosuppressive microenvironment [33, 61].

Moreover, nuclear ERα expression has been identified in lung adenocarcinoma and has been positively correlated with the presence of regulatory T lymphocytes (Treg), which exhibit the marker FoxP3+. The presence of Treg lymphocytes in the tumor area has been related with worse prognosis of NSCLC patients, due to these cells promote an immunosuppressive response allowing the tumor growth and metastases [62].

The chemokine-receptor CXCR4/CXCL12 pathway also promotes an immunosuppressive microenvironment in NSCLC. High CXCL12 expression by tumor in lung adenocarcinoma was associated with an increase of inflammation and CD4+ CD69+ CXCR4+ infiltrated lymphocytes, which 30% were Treg lymphocytes with CD25+ and FoxP3+ markers [63]. Previously we reported that CXCR4/CXCL12 pathway activation could be stimulated by estrogen. Expression of both, CXCR4 and CXCL12, chemokine and its receptor respectively, was higher in premenopausal women compared to postmenopausal women and men, suggesting an association with sexual hormones mainly with estrogens [14]. Furthermore, we showed that estrogen treatment stimulated in a dose- and time-dependent manner, the CXCR4 expression in lung cancer cell lines. Migration dependent of CXCR4/CXCL12 pathway in lung cancer cells was also observed after estrogen treatment [64]. These data suggest that estrogen could support an immunosuppressive response also through CXCR4/CXCL12 pathway and could promote other events in lung carcinogenesis, such as proliferation, cell migration, angiogenesis, metastases, epithelial mesenchyme transition and chemoresistance, by activating this chemokine/receptor pathway.

On the other hand, in hormone dependent cancer such as breast and endometrial cancer, estrogen also regulates the expression of the death protein ligand-1 (PD-L1) by tumor cells; this protein blocks the cytotoxic T lymphocytes activity binding its PD-1 receptor. This is a mechanism by which tumor cells evade immune system. Due to the above, immunotherapy based on blockade of the PD-1/PD-L1 pathway is currently, a new therapeutic strategy in NSCLC treatment.

Although the direct relationship between estrogen and PD-L1 in lung cancer has not been identified, it has been reported that PD-L1 expression is associated to EGF/EGFR activation pathway in tumors from NSCLC with EGFR mutational profile [65] and as we mentioned previously, EGFR pathway is strongly influenced by estrogen in NSCLC.

Furthermore, the relationship between PD-L1 expression and CXCR4/CXCL12 pathway activation has been identified in cellular hepatocarcinoma. CXCR4/CXCL12 activation under hypoxia condition, stimulated PD-L1 expression by tumor cells and the recruitment of Treg lymphocytes, while the treatment with AMD3100 a CXCR4 antagonist suppressed this polarization to immunosuppressive microenvironment [66]. We previously identified that CXCR4/CXCL12 expression, is associated with hormonal status of the patient and the pathway activation is stimulated by estrogen in vitro [14, 64], thus estrogen could establish an immunosuppressive tumor microenvironment through CXCR4, EGFR and PD-L1 expression.

ANTIESTROGENIC THERAPY IN NSCLC TREATMENT

Due to all the mechanisms by which estrogen stimulates lung carcinogenesis and the impact it has on the prognosis and survival of the patients, the estrogenic pathway has recently been considered as an important target for development of new therapeutic strategies in NSCLC treatment. There are several studies *in vivo* and *in vitro* that have test the effectiveness of antiestrogen drugs in NSCLC (Table 1). Due to the results of these studies currently there are 10 clinical trials investigating the toxicity, tolerated doses, and effectiveness of cancer therapies targeting estrogen signaling.

Table 1. Findings of preclinical and clinical trials testing antiestrogen drugs in NSCLC [79]

Antiestrogen drug		Results of preclinical studies	Results of clinical trials
ARO inhibitors	Anastrozole	Decrease of tumor growth	In process
	Exemestane	Decrease of tumor cell proliferation, migration and invasion. Increase in apoptosis of tumor cells	In process
	Letrozole	Reduction of proliferation	Toxicity grade 5 administered with everolimus
	Dexamethasone	Tumor growth reduction	-
ER down-regulators	Fulvestrant	Tumor growth reduction, decrease in mesenchymal features, increased sensitivity to chemotherapy and lysis by immune cells	In process
SERMs	Tamoxifen	Decrease of cell migration	Improved effectiveness of chemotherapy based on platinum and docetaxel
	Toremifene	-	Potentiated platinum toxicity
Combined treatment	Fulvestrant/ Gefitinib	Decrease of cell proliferation and tumor growth, increase in apoptosis cell death, increased sensitivity to gefitinib. Decreased cell proliferation in resistance mutation	Anti-tumor activity in pretreated female patients with advanced or recurrent NSCLC Increase in overall survival
	Fulvestrant/ Erlotinib	-	Enhances clinical benefits compared to separated therapy

Aromatase Inhibitors

There are two classes of aromatase inhibitors: 1) irreversible steroidal inhibitors, which block aromatase activity binding strongly to enzyme, these include drugs such as exemestane; 2) non-steroidal inhibitors, which reduce estrogen synthesis by reversible competition with the activity domain of aromatase enzyme, and include drugs such as anastrozole and letrozole.

Exemestane has showed suppressed tumor growth, decreased proliferation and increased apoptosis in tumors with high aromatase expression [67]. Also, decreased cell migration and invasion was observed in lung tumor cells treated with exemestane [68]. Anastrozole treatment also showed reduced tumor growth *in vitro* and *in vivo* models, while letrozole treatment reduces cell proliferation *in vitro* [19]. However, a clinical trial showed that the exemestane and everolimus (mTOR inhibitor) combined treatment caused toxicity grade 4 to 5 in patients with NSCL [69].

Estrogen production in tissues is controlled *in situ* by aromatase and enzymes that regulate its metabolism such as estrogen sulfotranferase (EST). Activation of EST by dexamethasone in A549 lung cancer cells results in a decreased level of estrogen and suppression tumor growth. Dexamethasone, one of the most widely used synthetic glucocorticoids, has shown promising anticancer efficacy in lung cancer, however further preclinical studies are needed to support its results [70].

Most studies have been conducted to test the efficacy of estrogen receptor inhibitors in NSCLC. Fulvestrant is a selective estrogen receptor degraded (SERD), it binds to ER inducing its destabilization and consequently its degradation. Fulvestrant has been tested in NSCLC preclinical studies, which have shown approximately 32% of tumor growth inhibition in mice previously treated with estradiol [71], also decreased ERβ expression has been noted [72]. Due to the relationship between estrogen and other signaling pathways in lung cancer, the efficacy of combined drugs has been investigated. The co-treatment with fulvestrant and vandetanib (a multi-target inhibitor of VEGFR and EGFR) synergistically decreased tumor growth and increased apoptosis significantly; fulvestrant enhanced the anti-tumor effects of vandetanib in NSCLC [56]. Moreover, fulvestrant treatment decreased mesenchymal features and increased tumor sensitivity to chemotherapy and immune activity of T lymphocytes [73].

Other important estrogen pathway inhibitors include the selective estrogen-receptor response modulators (SERMs). Within this group of inhibitors are drugs such as tamoxifen, raloxifen, toremifene, between others. The effect of tamoxifen in preclinical studies has been investigated in NSCLC. Tamoxifen, enhance the effectiveness of platinum chemotherapy and overcome platinum resistance through PKC inhibition, which in turn downregulates c-FOS expression [74]. In addition, tamoxifen decreased estrogen induced CXCR4 expression in lung adenocarcinoma cell lines, also a reduced migration of these cells was observed after tamoxifen treatment [64].

Interestingly, the treatment with fulvestrant in patients with breast cancer reduced the risk of developing additional primary malignant lung tumors. Clinical trials in phases I and II have showed that tamoxifen and toremifene treatment improved platinum toxicity in NSCLC, doses of toremifene, tamoxifen and cisplatin had good efficacy with safe and tolerable profile in patients previously treated with platinum compounds [74, 75]. However, other study shows no significant differences in the response or survival rates in NSCLC patients treated with toremifene and platinum chemotherapy [76].

On the other hand, the efficacy and safety of combined treatment with docetaxel and tamoxifen has been evaluated recently in a clinical trial, in this study NSCLC patients received the first-line platinum-based chemotherapy, and the results showed that the overall response rate and disease control rate in tamoxifen/docetaxel treatment were significantly higher than those in docetaxel group. However, the use of SERMs in lung cancer treatment has showed contradictory effects due to these compounds can act as both agonist and antagonist on ERs. Further preclinical studies are necessary before SERMs can be used in clinical trials exploring their therapeutic potential in ER positive NSCLC patients.

Due to the relationship between EGFR and ER signaling pathways, the combined treatment using inhibitors of tyrosine kinase (TKI) that block the EGFR pathway and antiestrogen drugs has been tested in clinical trials. The randomized multicenter phase II clinical trial TORI L-03, that assess the efficacy of combined erlotinib plus fulvestrant treatment in NSCLC, reported that patients with *EGFR* wild type tumors treated with this combination exhibited a higher clinical benefit rate compared to patients treated with erlotinib only [77].

A very interesting report from Collins and coworkers (2010) showed the clinical case of a female patient diagnosed with ER+, PR-, TTF-1+, EGFR+ and stage IV lung adenocarcinoma, treated with chemotherapy in combination with cetuximab (a EGFR monoclonal antibody) and bisphosphonate. The patient showed partial response and subsequently tumor progression. After 3 weeks of exemestane treatment (an aromatase inhibitor), the symptoms of the patient significantly improved and a 58% of decreased size of tumor was noted. The partial response was maintained for 6 months [78].

Currently the treatment of NSCLC has focused on immunotherapy, which blocks the immunosuppression mechanisms of the tumor and stimulates the immune response against the tumor. Immunotherapy based on blocking PD-1/PD-L1 pathway in lung cancer has significantly improved the overall survival of patients, however results of a meta-analysis, recently showed that the response to immunotherapy based on blocking PD-L1 was different by sex, being higher in male NSCLC patients compared to females. Although the effect of hormonal status (premenopausal/postmenopausal) on the response to treatment was not evaluated, this study shows that PD-L1 expression, as well as the successful of immunotherapy can depend on sex of patient and to be influenced by sexual hormones, mainly estrogens [11]. However the relation between estrogen pathway, the immune response to tumor and the immunotherapy in lung cancer is an area that must be explored.

Taken together these findings show that anti-hormonal therapy could be a new promising alternative for the treatment of lung cancer.

CONCLUSION

Several studies have supported the role of estrogen in lung carcinogenesis and tumor progression. ERβ and aromatase enzyme expression in NSCLC have been considered as important prognostic factors associated with worse prognosis and poor survival in NSCLC patients, thus estrogen pathway has emerged as a new potential target for the treatment of NSCLC.

Antiestrogen therapy has been used in the treatment of other hormonal cancers such as breast cancer and it is possible that it will be soon introduced in lung cancer treatment. However, it is necessary to establish a hormonal profile to identify patients whose tumors express ER+ and ARO+ and can be candidates for antiestrogenic therapy. Moreover, antiestrogen therapy should have a greater impact than the current target therapies, and it is estimated that it may improve the treatment of lung cancer since up to 80% of tumors from women and men who overexpress the ERβ and the aromatase enzyme.

Nowadays, there are ten clinical trials studying the effect of antiestrogen therapy, the findings of some trials have not published, however results of clinical trials, mainly those that tested the combined TKI and antiestrogen drugs, showed an incremented survival rate which suggests that antiestrogen therapy could benefit NSCLC patients.

In addition, due to the relationship between estrogen and EGF/EGFR, PD-1/PD-L1, VEGF/VEGFR pathways, the antiestrogen drugs in combined therapy, could also decrease the possibility of resistance due to the relationship between this signaling pathways, however further studies are needed to support this hypothesis.

Finally, the way in which estrogens stimulate carcinogenesis and tumor progression is a highly complex process and depends on several factors such as the concentration of the hormone, the expression of the receptors and its cellular location, as well as their relationship with different signaling pathways. Because of that, it is important to strengthen the research on new mechanisms of these hormones in lung cancer and in identifying the relationship with other pathways to recognize new therapeutic targets, and finally to understand the conditions by hormonal therapy could be a benefit for patients with lung cancer.

REFERENCES

[1] Globocan, 2018. *Global Cancer Burden in 2018.*

[2] Eilstein et al. *Cancer Epidemiol.* 2012; 36, e341-348.

[3] Islami F, et al. *Transl Lung Cancer Res.* 2015; 4,327-338.

[4] Herbst RS, et al. *N Engl J Med.* 2008; 359, 1367-1380.

[5] Wong MCS, et al. *Sci Rep.* 2017; 7, 14300.

[6] Vidyullatha P. *Gene Reports* 2016; 4, 198–207.

[7] Papadopoulos A, et al. *Br J Cancer.* 2014; 110, 1385-1391.
[8] Rosell R, et al. *N Engl J Med.* 2009; 361, 958-967.
[9] Barrera-Rodriguez R, et al. *Lung Cancer (Auckl).* 2012; 3, 79-89.
[10] Isla, D, et al. *Clin Transl Oncol.* 2017; 19, 527-535.
[11] Conforti, F, et al. *Lancet Oncol.* 2018; 19, 737-736.
[12] Hsu LH, et al. *Cancer Sci.* 2015; 106, 51-59.
[13] Moore KA, et al. *Ann Thorac Surg.* 2003; 76, 1789-1795.
[14] Rodriguez-Lara V, et al. *Arch Med Res.* 2014; 45, 158-169.
[15] Rodriguez-Lara V, et al. *Salud Publica Mex.* 2019; 61, 265-275.
[16] Rodriguez-Lara V, et al. *Nova science publishers* 2012; 149-163.
[17] Tam A, et al. *BMC Womens Health.* 2011; 11, 24.
[18] Massaro D, et al. *Am J Physiol Lung Cell Mol Physiol.* 2004; 287, L1154-L1159.
[19] Weinberg OK, et al. *Cancer Res.* 2005; 65, 11287-11291.
[20] Skjefstad K, et al. *Steroids.* 2016, 113, 5-13.
[21] Niikawa H, et al. *Clin Cancer Res.* 2008; 14, 4417-4426.
[22] Oton AB, et al. *J Clin Oncol.* 2006; 24, 7038.
[23] Ross H OF, et al. *J Clin Oncol.* 2007; 25, 7683.
[24] Schabath MB, et al. *Clin Cancer Res.* 2004; 10, 113-123.
[25] Brinton LA, et al. *Cancer Causes Control.* 2012; 23, 487-496.
[26] Li W, et al. *Steroids.* 2017; 118, 47-54.
[27] Adami HO, et al. *Int J Cancer.* 1989; 44, 833-839.
[28] Slatore CG, et al. *J Clin Oncol.* 2010; 28, 1540-1546.
[29] Greiser CM, et al. *Maturitas.* 2010; 65, 198-204.
[30] Chlebowski RT, et al. *Lancet.* 2009; 374, 1243-1251.
[31] Ganti AK, et al. *J Clin Oncol.* 2006; 24, 59-63.
[32] Chlebowski RT, et al. *Clin Lung Cancer.* 2016; 17, 10-17.
[33] Siegfried JM, et al. *Semin Oncol.* 2014; 41, 5-16.
[34] Chakraborty S, et al. *Expert Rev Respir Med.* 2010; 4, 509-518.
[35] Hershberger PA, et al. *J Steroid Biochem Mol Biol.* 2009; 116, 102-109.
[36] Mah V, et al. *Lung Cancer.* 2011; 74, 318-325.
[37] Kerr A, et al. *Adv Exp Med Biol.* 2008; 617, 377-384.
[38] Lund-Iversen M, et al. *Anticancer Res.* 2018; 38, 2261-2269.

[39] Leung YK, et al. *Proc Natl Acad Sci USA*. 2006; 103, 13162-13167.
[40] Stabile LP, et al. *Clin Cancer Res.* 2011; 17, 154-164.
[41] Jala VR, et al. *BMC Cancer.* 2012; 12, 624.
[42] Slowikowski BK, et al. *Mol Biol Rep.* 2017; 44, 35-50.
[43] Liu C, et al. *J Med Oncol.* 2015; 32, 1–13.
[44] Zhao G, et al. *J Steroid Biochem Mol Biol.* 2011; 124, 47-57.
[45] Marquez-Garban DC, et al. *Steroids.* 2011; 76, 910-20.
[46] Hershberger PA, et al. *Cancer Res.* 2005; 65, 1598-605.
[47] Marquez-Garban DC, et al. *Ann N Y Acad Sci.* 2009; 1155, 194-205.
[48] Stabile LP, et al. *Cancer Res.* 2002; 62, 2141-50.
[49] Bethune G, et al. *J Thorac Dis.* 2010; 2, 48-51.
[50] Levin ER. *Mol Endocrinol.* 2003; 17, 309-317.
[51] Giovannini M, et al. *J Thorac Oncol.* 2008; 3, 684-685.
[52] Nose N, et al. *J Clin Oncol.* 2009; 27, 411-417.
[53] Kawai H. *World J Clin Oncol.* 2014; 5, 1020-1027.
[54] Xie Q, et al. *Nan Fang Yi Ke Da Xue Xue Bao.* 2015; 35, 98-102.
[55] Zhang G, et al. *Mol Endocrinol.* 2010; 24, 1737-1747.
[56] Siegfried JM, et al. *J Thorac Oncol.* 2012; 7, 485-495.
[57] Siegfried JM, et al. *Oncotarget.* 2017; 8, 24063-24076.
[58] Tang H, et al. *Med Oncol.* 2012; 29, 2640-2648.
[59] Nurwidya F, et al. *Malays J Med Sci.* 2016; 23, 9-21.
[60] Rothenberger NJ, et al. *Int J Mol Sci.* 2018; 19(2).
[61] Navarro Flor C, et al. *Gender and the Genome.* 2018; 2, 81-91.
[62] Kadota K, et al. *Oncotarget.* 2015; 6, 27505-27518.
[63] Wald O, et al. *J Immunol.* 2006; 177, 6983-6990.
[64] Rodriguez-Lara V, et al. *Endocr Res.* 2017; 42, 219-231.
[65] Chen N, et al. *J Thorac Oncol.* 2015; 10, 910-923.
[66] Chen Y, et al. *Hepatology.* 2015; 61, 1591-1602.
[67] Koutras A, et al. *Mol Cancer.* 2009; 8, 109.
[68] Giannopoulou E, et al. *Biochim Biophys Acta.* 2015; 1853, 328-337.
[69] Singhal N, et al. *Cancer Chemother Pharmacol.* 2015; 75, 325-331.
[70] Wang LJ, et al. *Acta Pharmacol Sin.* 2016; 37, 845-856.
[71] Stabile LP, et al. *Cancer Res.* 2005; 65, 1459-1470.
[72] Tang H, et al. *Oncol Res.* 2014; 22, 13-20.

[73] Hamilton DH, et al. *Clin Cancer Res.* 2016; 22, 6204-6216.

[74] Perez EA, et al. *Cancer Invest.* 2003; 21, 1-6.

[75] Lara PN Jr, et al. *Cancer Chemother Pharmacol.* 2001; 48, 22-28.

[76] Cheng H, et al. *Zhongguo Fei Ai Za Zhi.* 2003; 6, 335-338.

[77] Garon ES, et al. Results of TORI-L-03, a randomized, multicenter phase II clinical trial of erlotinib (E) or E + fulvestrant (F) in previously treated advanced nonsmall cell lung cancer (NSCLC) in *Proceedings of the 104th Annual Meeting of the American.* 2013. Washington DC. Philadelphia.

[78] Collins IM, et al. *J Thorac Oncol.* 2010; 5, 749-750.

[79] Rodriguez-Lara V, et al. *J Thorac Dis.* 2018; 10, 482-497.

In: Estrogen Receptors
Editors: George G Chen et al.
ISBN: 978-1-53618-228-6

Chapter 6

ESTROGEN RECEPTOR STRUCTURE: EXPRESSION AND ROLE IN GENDER DIFFERENCES OF AIRWAY DISEASES

Olga Carvalho**, *PhD

Institute of Histology and Embryology, Faculty of Medicine,
University of Coimbra, Coimbra, Portugal

ABSTRACT

Estrogen's biological effects in several organs are mediated by two receptors ER alpha and ER beta. These receptors have similar structures but exhibit distinct functions in diverse organs as well as different tissue distributions.

In the past years gender differences in airways diseases have been observed, and sex hormones play an important role in the development and outcome of some respiratory pathologies.

In this review, I characterized estrogen receptors and searched for a possible correlation between estrogen and gender disparities in airway diseases.

Keywords: estrogens receptors, airway diseases, lung; gender differences

---

* Corresponding Author's Email: omcarvalho@gmail.com.

INTRODUCTION

There is increasing clinical epidemiologic data that confirm gender differences in the incidence, prevalence, morbidity, and mortality of airway diseases including asthma, chronic obstructive pulmonary disease (COPD), cystic fibrosis (CF), lung cancer and non-CF-related bronchiectasis.

Although these conditions have distinct pathophysiologies, women for largely unknown reasons are increasingly becoming more prevalent and experiencing excess morbidity and mortality from these disorders, suggesting a higher risk factor for female gender in inflammatory lung diseases.

The gender dichotomy renewed the interest in researchers, and the exact reason for these differences in human health and disease is probably multifactorial. Nevertheless, genetic inherent and/or steroids hormones play important roles because these two rule host physiology, immunity, microbiota and psychological or social behavior.

Steroids hormones regulate innumerous aspects in vertebrates and estrogens, can influence the growth, differentiation and functioning of male and female reproductive systems organs, and also play an important role in bone maintenance and in cardiovascular system. They are mainly produced in the ovaries and testis, and then diffused in and out of all cells, being retained in the specific target cells by a nuclear receptor protein, termed the estrogen receptor (ER), that controls the gene expression.

The biological effects of estrogen occur through the activation of an estrogen receptor alpha (ERα) or/and beta (ERβ), which have highly homologous strutures but non-identical tissue distribution and action [15].

And since the emerging data suggests that female sex hormones play a role in inflammatory airway disesases, we will describe estrogen receptor struture, expression and examine the potential role of estrogen in the pathogenesis and developing of airway disease.

ESTROGEN RECEPTORS

The biological effects of estrogen are expressed through the activation of two estrogens receptors (ERs) [55], the first ERα receptor to be cloned and characterized [38] and ERβ, which is highly homologous with the "classical" ERα and has an overlapping but non-identical tissue distribution [77].

Estrogens receptors, belong to the superfamily of nuclear receptors, and have an important role in several processes including regulation of reproductive and cardiovascular system, bone density, strength and regeneration of muscle [47] and inflammation [37; 75].

The nuclear receptors are ligand-activated transcription factors that regulate the expression of target genes by binding to specific acting sequences [107]. This superfamily comprises the steroids, retinoids, thyroid and vitamin D3 receptors [28], and can be broadly divided into four classes based on their dimerization and DNA- binding properties [107].

The typical nuclear receptor structure presents a variable amino-terminal region (A/B), a highly conserved DNA-binding domain (C), a variable hinge region (D), a conserved carboxyl-terminal ligand binding domain (E), and a variable carboxyl-terminal region (F) [52; 107] (Figure 1).

Figure 1. Nuclear receptors structure [15].

The N-terminal (A/B) contributes to the transactivation function; the C domain is responsible for specific DNA binding domain (DBD), dimerization and localization, and the carboxyl-terminal region is involved in ligand-binding, nuclear localization, and ligand-dependent transactivation function. The carboxyl-terminal half of the receptor encompasses the ligand-binding domain (LBD), which possesses the essential property of hormone recognition and ensures both specificity and selectivity of the physiologic response [52; 107]. In simplest terms, the LBD can be thought as a molecular switch that, upon binding ligand, shifts the receptor to a transcriptionally active state [123].

When the nuclear receptors bind to ligand, it causes a change in the receptor conformation that result in dimerization, and binding to a specific promoter sequence of DNA, called hormone response elements (HREs). The activated receptor/DNA complex recruits others cofactors from the nucleus, which results in transcription of DNA downstream from the HRE in to mRNA and causes a change in cell function [107].

ESTROGEN RECEPTORS A AND B STRUCTURE

The physiological effects of estrogens are manifested through two ER isoforms – ERα and ERβ – which display distinct regions sequences [78; 85]. In human, two domains are essential to activate efficiently the transcription of estrogen-responsive genes, the hormone binding (E) and tight nuclear binding (C). Region D can be altered without affecting activation, and deletion of the N-terminal domain (A/B) has no effect on activation of a reporter gene [53].

All domains of the structural architecture of ERα and ERβ receptors are similar, but their homology depends on the composition of each domain (Figure 2).

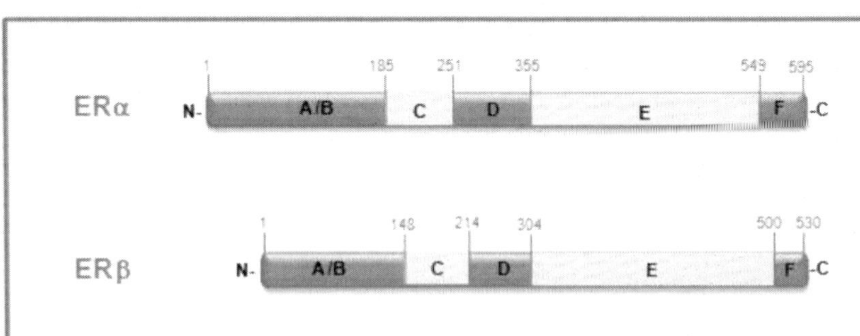

Figure 2. Human estrogens receptors [15].

In the two human ERs, the LBD is relatively well conserved (58%) which suggests that there is a high degree of structural conservation of the ligand-binding pocket(s), but different ligand affinity exists [49; 77]. Although the DNA-binding (C domain) is extremely well conserved (96%), ERβ and ERα may activate different sets of target genes [77]. The A/B domain, the hinge (D) and the F domain are poorly conserved. The lack of conservation of the A/B domain suggests important differences between both receptors with respect to their activation function (AF-1) activity and with respect to interactions with co-activators and/or repressors in this region.

There are two activation domains in ERα, an N-terminal ligand-independent activation function (AF-1) and a C-terminal ligand-dependent activation function (AF-2) [108]. These act synergistically to recruit various coactivator proteins to the DNA/ER complex. Depending on cellular and promoter context, when the coactivator bind exclusively to AF-1, occurs either partial or no activation of ERα [118; 73]. For full activation of the receptor, generally is required that the coactivators bind simultaneously to both AF-1 and AF-2 [5]. In contrast, ERβ appears to have diminished AF-1 activity while, at the same time, it possesses a fully functional AF-2 [26; 69].

The LBD three-dimensional structures of both ERs have high sequence identity and in both cases, the polypeptide chain is folded into the canonical α-helical sandwich topology associated with NR-LBDs [82]. This compact "wedge-shaped" domain is composed of 12 helices (H1–H12) that are arranged into three anti-parallel layers. The observed helical arrangement acts as a molecular scaffold that maintains a sizeable buried cavity at one end of the LBD [82].

Although most regions of the LBD involved in ligand interaction were observed in both ERs isoforms, ERβ has a different degree of plasticity in its LBD cavity when compared with ERα, which indicates that the LBD of ERβ undergoes a different structural response to the same ligand when compared to ERα [23]. This also suggests distinct receptor conformational changes that may lead to unique patterns of coregulator interaction and, ultimately, tissue selective pharmacological outcomes [23; 88].

Both ERs have a second binding site in the ligand binding domain (LBD) but these second site seems to have no obvious function [119]. Probably, the second binding site may be an evolutionary remnant, which is in accordance with his apparent lack of function and with the absence of the binding site in other nuclear receptors, such as retinoid RARγ [119]. Nevertheless, the estrogen antagonist 4-hydroxytamoxifen (HT) can occupy not only the core binding pocket within the ligand-binding domain of estrogen receptor ERβ but also bind to the surface of the second site [98; 122].

It is believed that agonist or antagonist binding induces a distinct conformation change upon the receptor which initiates activation or suppression of responsive genes. Classic ERs binding models were validated by the different ER/ligand complex structures in which ligands reside in the steroid binding site and interact with ER-LBD to modulate the position of helix 12 in distinct positions either favouring or disrupting coactivator binding [11; 83; 98].

A wide list of structurally distinct compounds binds to ERs with differing degrees of affinity and potency [50]. Some of these compounds, such as 17β-estradiol (E2) are natural ligands that act solely as agonists, whereas others function as pure antagonists. A third category, termed as selective ER modulators (SERMS), have the ability to act as both agonists and antagonists depending on the cellular and promoter context, as well as the targeted ERs isoform [39; 62; 80].

The ligand binding affinity of ERα and ERβ protein is overall similar for the physiological ligands, and the most interesting difference was found for 17α-estradiol (E2), which has a five times higher affinity for ERα. In fact, E2 is a short-acting estrogen, as well as estrone and estriol that can cause nuclear binding of the hormone-receptor complex, but only for a short period of time [50]. Early studies showed that E2 is the most potent endogenous ligand for ERβ, and that it binds equally well to both ERs [41].

Besides the ERs transduction, another protein, G protein-coupled receptor (GPR30) has been recently suggested to transduce some of the estrogenic responses [89].

FETAL ESTROGEN RECEPTORS EXPRESSION

Estrogens physiological activity begins during fetal development, when ERs are detected in the blastocyst stage, suggesting a possible functional requirement of estrogen in embryos preimplantation [34; 127]. During the last trimester of fetal life, both ERs are expressed in the human fetal ovary, which suggests that estrogen action already exists and that may have a role in the ovary developing after follicle formation [120].

In the human fetus, high levels of ERα mRNA were detected in uterine tissue, and a weaker signal was associated with ovary, testis, skin, and stomach tissues [9]. The ERβ mRNA exhibits strong signal in the ovary, testis, spleen and adrenal glands of human fetus, a moderate signal in thymus, uterus, brain cortex, pituitary kidney, skin, and lung; and a weaker expression was found in the gut, muscle, heart, and placenta [9].

Nevertheless, in another study with human fetal tissues, ERα mRNA was undetectable or present in very low levels between developmental gestation weeks 13-20, but ERβ mRNA was detected in various tissues [103]. The highest level of ERβ expression was found in adrenal glands, male and female reproductive tracts, and in low levels, in the brain, heart, lung and kidney [103].

In the mice, ERs are present in male and female reproductive tissues from gestation day E10 and later, including the gonads and primitive reproductive tracts [35; 36]. During development, and depending on the ERs isoform, estrogen receptor can also be expressed in the atrial wall, brain, larynx, kidney, urethra, bladder mammary gland, and midgut [58].

The study of mice ER-deficient (ERKO) showed the importance of both receptors for the normal development of lung. ERα ensures that the lungs differentiate properly; leading to normal numbers of alveoli per surface area, and ERβ modulates the development of extracellular matrix, leading to normal elastic tissue recoil pressure in lungs [64; 65].

In the mice´s lung, no ERα was detected during gestation days E15 to E19, which suggested that throughout this development period the estrogenic action is mediated predominantly by the ERβ receptor [14]. In fact, the ERα knockout mice suggest that ERβ may substitute the ERα during fetal development [61; 40].

There has also been observed a gender difference in the expression of the ERβ in the fetal mice lung. Female lungs tissues expressed ERβ from developmental day E15 to E19, while the male lung tissues only express this receptor from developmental day E16 to E18 [14]. During this period of development, female fetal lungs are exposed to a longer mediation of estrogens action via ERβ when compared to male mice and this difference may be important for the lung development [14].

ADULT ESTROGEN RECEPTORS EXPRESSION

ERα and ERβ are products of different genes and exhibit different tissue expression [50], divergent tissues response and distinct physiological and pathological activities [22; 43].

Human ERα receptor has been immunolocalised in several organs such as Fallopian tubes, ovary, cervix, uterus, testis, bladder, pituitary and others [106]. ERβ was localized in a wide range of normal adult tissues including ovary, Fallopian tube, uterus, kidney, brain, heart, testis and thymus [77; 106]. Only ERβ is expressed in tissues that are not considered as sensitive to estrogens, such as lung epithelium [81], ventral prostate [124] and granulosa cells [20].

In the adult rat, the highest levels of ERβ mRNA were reported in the ovarian follicle granulosa cells and in the prostate epithelium [50; 51]. The testis, uterus, bladder, and lung showed moderate expression, whereas pituitary, epididymis, thymus, thalamus/hypothalamus, and spinal cord reveal low expression of ERβ mRNA [50]. The ERα mRNA was highly expressed in ovary, uterus, epididymis, testis, pituitary, uterus, kidney, and adrenal [50].

The study of mice with knockout of ERα, or ERβ, or both receptors revealed that ERs isoforms have overlapping but also unique roles in estrogens-dependent action in vivo. Each receptor has different transcriptional activities in certain ligand, cell-type, and promoter contexts, but can be coexpressed in a number of tissues [69]. When both receptors are expressed in the same tissue, ERα promotes the proliferation while ERβ has pro-apoptotic and pro-differentiating functions [75]. One important characteristic of ERβ seems to be the capacity to modulate the biological activity of ERα in the uterus [125].

Apparently, the ERβ is involved in important physiological processes such as differentiation, extracellular matrix organization and stromal-epithelial communication [49]. Present data suggest that ERβ plays a minor role in mediating the estrogen action in several organs/organ systems consider as estrogen target tissues, such as uterus or normal mouse mammary gland development, and is unable to influence hypothalamus/ pituitary axis [41]. Nevertheless, ERβ has a role in folliculogenesis [41].

Human lung tissue expresses both ERs mRNA, but ERα transcription is relatively less abundant than ERβ, and the ERβ protein expression is twice of the ERα [42].

In the adult mouse ERβ mRNA is the only isoform expressed in the lung. The highest levels of ERβ mRNA were detected in lung of both genders [22] and are detected in the ciliated and non-ciliated cells of bronchial epithelium and in the alveolar cells type I and type II [75; 81]. This localization indicated that ERβ should mediate the estrogens effects in the lung and maybe responsible for the female's pulmonary phenotype [65; 66]. In fact, a gender difference has been observed in the rat's lungs during post maturation, with adult virgin females having smaller but higher numbers of alveoli and bigger alveolar surface area compared with aged-matched males [64; 67].

Several animal studies indicate that estrogens are required for the formation and complete development of the alveoli [64; 65; 113], the maintenance of existing alveoli, the lung homeostasis [81] and the induction of alveolar regeneration in adult mice [65; 66].

The functional disruption of ERα in mice leads to changes in a variety of respiratory parameters, which suggest that this nuclear receptor may be a critical regulator of breathing and respiratory rhythmogenesis [1; 12].

There seems to be a difference in the translation of estrogen signal, in the case of lung development and lung regeneration. For the regeneration it seems that is necessary the receptor ERα but little or none interference of ERβ is required, while for the full development of the alveoli the two receptors are essential [66].

Via ERβ, the estrogens also modify the lung expression of PDGF-A (factor platelet-derived growth), which in turn alters the proliferation, migration and deposition of elastin, thus influencing alveologenese. They can also modify the expression of GM-CSF (granulocyte colony-stimulating factor-macrophages), which influences the action of surfactant homeostasis [81].

AIRWAY DISEASES AND GENDER DIFFERENCES

Bronchiectasis is a complex multifactorial chronic respiratory disease characterized by an abnormal, permanent and irreversible dilatation in the conducting bronchial airways. This disease is the result of a "vicious cycle" of chronic inflammation and infection that leads to airways dilatation, thinning of capillary walls, and impaired mucociliary clearance [8]. The patients have persistent airway inflammation, chronic cough and excessive mucopurulent secretions [8]. The main cause of bronchiectasis is the genetic disease cystic fibrosis (CF), but it may also be driven by a range of non-CF related disorders or be idiopathic [121].

Non-CF bronchiectasis has highly variable clinical presentations that sometimes overlap with other respiratory disease states. In response to pulmonary infection or tissue injury, the predisposed individual develops a robust inflammatory response that is partially responsible for the structural damage of the airways. The structural abnormalities allow for mucus stasis, which favors continued chronic infection. Repeated cycles of infection and inflammation coupled to mucus hypersecretion leads to obstruction and collapse of smaller airways further perpetuating more bronchiectasis [71].

In non-CF bronchiectasis, ender-related differences are described in disease prevalence, severity and outcome. The prevalence is higher in males but the disease is more severe in females [60; 76]. Poorer clinical outcomes, worse lung function and a survival disadvantage are also reported in females when compared to males across all age groups [71; 87]. In the population under 18 years, while scarce data, the ratio of bronchiectasis suggested is 2:1 male: female [87].

In the gender dichotomy of bronchiectasis are involved physical and biological mechanisms. These include comorbidities, pulmonary anatomy and physiology, chronic infection and inflammation, impaired host defenses and other environmental influences [84].

Sex-related differences in respiratory tract structure, function and microbiome composition predispose females to earlier infection [121]. Females have smaller lungs and therefore smaller conducting airways [13]. Goblet cells located at the surface of pseudostratified ciliated epithelia are responsible for mucus production and estrogen regulates MUC5B gene expression augmenting mucin production [105]. Through the estrogen receptors expressed on airway epithelial cells, estrogens directly regulate goblet cell expression and impact the post-translational modification of mucin, a key component of mucus [105]. Estrogen and progesterone, through functional regulation of ion transporters, also influence airway cilia beat frequency and function thereby affecting the optimal mucociliary clearance, which when impaired confers a susceptibility to microbial colonization and infection [93].

The gender differences in microbial colonization, infection and virulence can, at least in part, be attributed to differences in host immune response and bacterial pathogenicity; features influenced and regulated by sex steroid hormones [93]. The respiratory microbiome appears to have sex-specific signatures and is susceptible to a range of host immune and inflammatory consequences [18].

In resume, sex steroid hormones potentially play an important role in the pathophysiology of non-CF bronchiectasis through anatomical variation, regulation of lung function and altering microbiota composition, as well as influencing host immune and inflammatory response.

Cystic fibrosis (CF) is the most common life-shortening inherited disease in the Caucasian population, affecting approximately 75,000 individuals worldwide. The maintenance of such high incidence rates in most European populations can perhaps be explained by an evolutionary perspective advantage [16]. It was considered lethal during infancy and childhood, but in nowadays individuals with CF can survive until 40 years of age due to an early diagnosis and an aggressive therapeutic attitude [31].

CF is an autosomal recessive disorder caused by mutations in the cystic fibrosis transmembrane regulator gene (CFTR) that leads to a defect in the ion transport across the epithelia of organs with exocrine function [91]. The CFTR protein is an anion channel of primary importance that controls the normal movement of chloride (Cl-), sodium (Na+), water in and out of the cells in different parts of the body [30].

This multi-system disease affects lungs, pancreas, gastrointestinal and reproductive system and is implicated in chronic lung infection, malabsorption, salt loss syndrome, infertility and development of other numerous morbidities. It can also impair other organs, such as the hypothalamus, kidney, bone, and maybe implied in growth retardation, delayed pubertal onset, bone density modulation and susceptibility to renal calculi [17].

The foremost cause of morbidity and mortality in CF is the lung pathophysiology [25]. In the respiratory system the CFTR protein plays a major role in determining the airway surface mucus and therefore impairs the mucociliary clearance. The airways mucus becomes abnormally thick, dehydrated and viscous and more difficult to clear by the ciliary beating, creating an environment for bacteria proliferation and infections [48; 68; 92]. These defects in the airways' innate defence trigger a chain of events that includes mucus stasis and plugging, airway obstruction, infection and inflammation [30].

Several epidemiological studies have evaluated the frequency, prevalence and mortality in CF patients during several years and found that females have a lower survival rate and a worse prognosis when compared to males [93]. Nevertheless, two other studies showed no difference in the survival rate between CF females and males but both studies included young

CF patients (children and adolescents) and this might explain why the gender difference, which may likely occur after puberty, could not be observed [93].

It is clear that there is a CF female disadvantage in the prognosis, survival and morbidity, the called "CF gender gap" and the investigation has been focused on determining the potential crucial role of estrogen in the establishment of this gender gap [93; 102; 131].

It has been shown that estrogen may modulate ion transport in airway epithelial cells and that this can have beneficial or deleterious effects depending on the tissue compartment and on the physiology/ pathophysiology state of the airways [3]. In the lung, estrogens appear to affect Na+ reabsorption and fluid clearance, indicating a regulatory role for female sex hormones on the physiological conditions of respiratory system [93].

The normal airway epithelium is covered by an aqueous film termed the airway surface liquid (ASL) that is the main protective barrier against the entry of pathogens and is composed of periciliary layer and the mucus layer. In CF subjects, the ASL height is decreased when compared with normal epithelium. The effect of estrogens on the ASL of CF bronchial cells show that 17β-estradiol (E2) causes a further decrease in ASL height and this impaired ASL volume homeostasis effect of E2 is regulated by the estrogen receptor ERα [21]. Estrogens can also upregulate MUC5B gene expression in normal human airway epithelial cells, which is one of the major mucins in the human airway submucosal glands. Estradiol may have the potential to augment mucin production resulting in a reduced clearance airway surface [105].

In resume, the molecular basis for the CF gender gap involves a compromised in the innate immune response via E2-ERβ receptor signal transduction and a reduced airway surface liquid dynamics and ion transport dysfunction via E2-ERα receptor signal transduction [93].

Asthma is a chronic inflammatory disease characterised by the reversible narrowing of the airways (bronchoconstriction) in response to diverse stimuli. The airway hyper-responsiveness (AHR) is defined by three salient features: 1) the airway wall inflammation that involves activation of mast cells, infiltration of eosinophils, and increased of activated T-cells, that mediate allergic inflammation through the secretion of an array of cytokines; 2) mucus luminal obstruction caused by the hyper-secretion of bronchial mucous glands along with infiltration of inflammatory cells; 3) vasodilatation of bronchial microvasculature with increased vascular permeability and edema [29]. Chronic inflammation can lead to structural changes in the airways; including airway smooth muscle hypertrophy and hyperplasia [63].

Asthma can be triggered by repeated antigen exposure (allergic asthma) or by abnormal autonomic neural regulation of airway function (nonallergic asthma). It causes coughing, wheezing, episodic dyspnea, tightness in the chest and reversible airway obstruction [4]. Many phenotypes and endotypes of asthma exist, ranging from mild asthma with onset during childhood to severe asthma with later onset, making asthma a broad disease with different pathologies [126].

A gender disparity has been established in asthma and this discrepancy change throughout life [130]. Before puberty, girls have lower prevalence and severity of asthma than boys [54] but during adolescence, asthma prevalence and morbidity in males decline concurrent with an increase in females [33]. By adulthood, women have increased asthma prevalence compared to men [74], and women are three times more likely than men to be hospitalized for an asthma-related event [19]. This gender discrepancy is maintained until around the time of menopause, when a decrease in asthma prevalence is noted in women [111]. It has also been observed that asthma symptoms get worse during menstrual cycle [10] and decreased during pregnancy [95].

All these data suggest that sex hormones are important in asthma pathogenesis, but the mechanisms by which estrogen and/or androgen signaling regulate airway inflammation, mucus production, and airway hyper-reactivity are not fully elucidated [129].

Animal studies using genetic deletions of estrogens receptors and mutated androgen receptor have shown that estrogen signaling promotes and androgen signaling attenuates inflammation [129]. Estrogen modulates the immune function via its estrogens receptors, ERα and ERβ, but also by a G protein-coupled receptor 30 (GPR30), which are expressed by a large number of immune regulatory cells [27; 128]. Estrogen receptor knockout mice have showed that estrogen increased airway inflammation and that effect is mediated through ERα receptor [12].

Non-immunological mechanisms, such as gender differences in airway hyper-responsiveness (AHR), smooth muscle contractility, and mucus production, may also drive the gender differences in asthma. The increased basal AHR in male mice is potentially caused by fewer numbers of alveoli and decreased alveolar surface area compared to female mice [64]. Additionally, ERα and ERβ signaling are important for alveolar development since ERα and ERβ deficient mice have decreased alveoli [64].

Estrogen and progesterone are also important in mucus production and mucociliary clearance in the airway epithelial cells that express estrogen receptors, predominantly ERβ, and progesterone receptors [129].

Asthma also seems to be more common in the overweight and obese population [6] and the interaction between body mass index (BMI) and asthma is stronger in women than men, suggesting that female sex hormones may play a role in the increased prevalence of asthma among obese women. Estrogens may modulate the immune response and increase the risk of asthma [46].

Chronic obstructive pulmonary disease (COPD) was historically considered an illness of elderly men, which was correlated with the high prevalence of smoking among men. But actually this disease is commonly seen in women, and the number of women dying per year is now greater.

Cigarette smoking remains the most important risk factor for developing COPD, however, it is becoming increasingly recognized that never smokers may account more than one-third of all worldwide COPD cases, with women making up more than two thirds of the non-smokers with moderate-to-severe COPD [56; 94].

In general, women smoke less than men, which suggest that they may be more susceptible to developing COPD in response to cigarette smoke exposure. A large population study, evidences that women appear to have more severe COPD with early-onset disease (under 60 yr) and a greater susceptibility with lower tobacco exposure, suggesting that women may be more vulnerable to the damaging respiratory effects of cigarette smoke [57;

99]. Furthermore, women with severe COPD have a higher risk of hospitalization and death from respiratory failure and comorbidities [86].

The reasons for the greater propensity of women to developed COPD is unknown but several studies try to demonstrate the mechanisms behind these sex differences, that include factors like genetic predisposition, sex hormones and comorbidities [2].

Smoking patterns between men and women may differ, with men taking deeper inhalations of cigarette smoke, although the number of cigarettes smoked and the age of starting smoking are similar between the sexes. Metabolism of cigarette smoke may differ in women because of sex differences in the expression and activity of cytochrome P450 (CYP) enzymes. It is possible that the inflammatory response to cigarette smoke is different in females because of hormonal effects, although this has never been demonstrated. Animal experiments shown that estradiol upregulates CYP enzymes, making the female lungs more susceptible to oxidant damage in response to cigarette smoke. The increased CYP expression is related to increased levels of estradiol and increased metabolism of cigarette smoke to generate oxidants/oxidizers, suggesting that female sex hormone contributes to oxidative stress and greater airway injury [97; 105].

The airways of females are relatively smaller than those of males for the same lung volume, so there may be a greater concentration of tobacco smoke per unit area of small airway surface [72].

Experimental study shows that only in female, the smoke-induced COPD is characterized by small airway remodelling through dysregulation of antioxidant defenses and activation of transforming growth factor-β1 (TGF-β), resulting in airflow limitation. This mouse study show that these effects can be ameliorated by ovariectomy before smoke exposure and that the tamoxifen (an estrogens receptor antagonist) attenuated smoke-induced oxidative stress in the lungs, suggesting that estrogens may be involved in the dysregulation of the oxidant/TGF-β signalling axis and therefore, implicated in the disease process [104].

Compared with women, man had a higher risk of dying by respiratory causes, but the loss of the protective effect of estrogens can be visible during menopause by the increased incidence of COPD in non-smoking women, with an incidence as twice of men [90; 101].

Overall, women are more prevalent and have worse prognosis in COPD but the exact mechanism behind the gender differences is still uncertain. The studies show that sex hormones, in particular estrogen, play an important role in the modulating of inflammatory airway conditions but the biological mechanism needs better understanding.

Preterm Infants and Airway Diseases

During gestation, human fetuses are exposed to an increasing amount of both estradiol and progesterone [117]. Preterm infants are deprived of this hormones supply at a much earlier developmental stage and for this reason, it is reasonable to consider that the postnatal replacement of these hormones in extremely preterm may be important for the lung development.

The estradiol and progesterone replacement therapy have shown that can reduce the incidence of chronic lung disease [109; 112; 115; 116]. Postnatal estradiol treatment enhanced pulmonary function and lowered the need of ventilatory support, and may be an efficacious postnatal therapy to improve lung function in bronchopulmonary dysplasia [70]. The estradiol and progesterone treatment also increased the mRNA expression of vascular endothelial growth factor (VEGF) and surfactant proteins, so important for the prenatal lung development and postnatal lung function [114].

Infants born prematurely suffer major morbidity and mortality than infants born at term which shows that gestational age remains the key factor in the development of neonate diseases [44; 132]. Among the morbidities, the greatest contributor to infant mortality is the respiratory distress syndrome (RDS) and subsequent bronchopulmonary dysplasia (BPD) or chronic lung disease (CLD), with higher rate in infants born prematurely due to insufficient surfactant.

Despite the advent of antenatal corticosteroid administration and postnatal surfactant use, clinical studies support a gender gap in neonatal outcomes, with a higher rate of morbidity and mortality in males [79; 100;

110]. Preterm males have a higher incidence of RDS, pneumothorax, early respiratory support and subsequent bronchopulmonary displaysia (BPD) [110]. The high mortality in preterm males can be largely attributed to increased acute respiratory disorders, which is due in part, to male delayed lung maturation [45].

This gender disparity remains undetermined but must be a multifactorial explanation, in which structural, immunological and hormonal intervention plays an important role.

Male fetuses tend to weigh more at any given gestational age and thus tend to have more alveoli and alveolar surface area than gestational age-matched females [13]. However, surfactant production appears earlier in female lung development than in males [32]. This earlier presence of surfactant seems to prevent the early closure of female alveoli and small airways, which may contribute to the higher airflow and decreased resistance of the female respiratory system [110].

In fact, earlier surfactant production is mediated by female sex hormones [113] and the lung development maturation is stimulated by estrogens [64]. The deletion of estrogen ERβ in female mice shows changes in the lung development, with an increase in alveolar size and a decrease in alveolar surface area, making these mice look like their male counterparts [110]. Furthermore, androgens have showed to inhibit surfactant production by altering epidermal growth factor and transforming growth factor β1 [24].

Overall, the decrease in surfactant and lower gas exchange area differences can strongly impact the risk of developing RDS in preterm male neonates [110].

The postnatal respiratory support in premature neonates exposes the undeveloped lung to high concentrations of oxygen (hyperoxia) increasing oxidative stress in the lung and contributing to the development of BPD [7].

The high incidence of BPD in males is not fully understood, but one explanation could be the sexual dimorphism of surfactant synthesis, probably due to the deleterious effects of androgens in the lung developing male fetus [96].

Experimental study has showed that when exposed to hyperoxia, the neonatal male mice are more susceptible, have more inflammation in the lungs, impair alveolarization, impair pulmonary angiogenesis and higher mortality compared with similarly female mice [59]. Sex-specific differences in alveolarization, angiogenesis, and inflammation could explain the increased incidence of BPD in male premature neonates [59].

These underscored male "disadvantage" remain relatively undetermined but include several factors such as genetic, immunological and hormonal and should be considered in future clinical and experimental research.

CONCLUSION

The worldwide increasing number of women with the airway diseases, such as asthma, CF, COPD, renewed the research interests to understand the important biological gender differences on lung mechanisms diseases, in particular the biological role of sex hormones in modulating airway inflammation, mucus production and cigarette de-toxification and other processes relevant to COPD, asthma and CF.

In fact, the research studies show that estrogen modulates airway epithelial ion channel regulation, decreases airway surface liquid, up-regulates mucus production, changes cigarette smoke de-toxification increased susceptibility to oxidant damage by tobacco.

The recognizing of this gender disparity can be important for timely diagnosis, prognostic and therapeutic approaches, and also for more customized patient care, incorporating increased patient education and effective counseling about symptoms particularly surrounding, for example, puberty, pregnancy and menopause.

The more we know about the mechanisms behind gender disparities in airway diseases, the further it can lead us to enhanced interventions and better outcomes for both men and women.

Acknowledgment

I would like to thank the Director of Institute of Histology and Embryology, Professor Carlos Gonçalves, for the support and for giving me more time to devote myself to the writing of this chapter.

References

[1] Ambhore NS, et al. *Cell Mol. Biol.* 2019; 61, 469.
[2] Barnes PJ. *Am. J. Respir. Crit. Care Med.* 2016; 193, 813.
[3] Bastarache JA, et al. *J. Crit. Care* 2011; 26, 249.
[4] Bateman ED, et al. *Eur. Respir. J.* 2008; 31, 143.
[5] Benecke A, et al. *EMBO Rep.* 2000; 151.
[6] Beuther DA, et al. *Am. J. Respir. Crit. Care Med.* 2007; 175, 661.
[7] Bhandari V. *Semin. Fetal Neonatal Med.* 2010; 15, 223.
[8] Boyton RJ, et al. *Annu. Rev. Pathol.* 2016; 11, 523.
[9] Brandenberger AW, et al. *J. Clin. Endocrinol. Metab.* 1997; 82, 3509.
[10] Brenner BE, et al. *Thorax.* 2005; 60, 806.
[11] Brzozowski AM, et al. *Nature.* 1997; 389, 753.
[12] Carey MA, et al. *Am. J. Respir. Crit. Care Med.* 2007; 175, 126.
[13] Carey MA, et al. *Trends Endocrinol. Metab.* 2007; 18 (8), 308.
[14] Carvalho O, et al. *Anat. Histol. Embryol.* 2012; 41, 1.
[15] Carvalho O, et al. In: *Estrogens receptors. Mechanisms, structure and role in disease,* George Chen (editor), Nova Science Publishers Inc., 2012, pp.133-147.
[16] Carvalho O, et al. In: *The Oxford Handbook of Evolutionary Medicine,* Martin Brüne and Wulf Schiefenhövel (editors), Oxford University Press, UK, 2019, Chapter 12, pp. 487-530.
[17] Castellani C, et al. *Cell. Mol. Life Sci.* 2017; 74, 129.
[18] Chamekh M, et al. *Front Immunol.* 2017; 8, 1806.
[19] Chen Y, et al. *J. Clin. Epidemiol.* 2003; 56, 180.
[20] Cheng G, et al. *J. Clin. Invest.* 2008; 118, 4025.

[21] Coakley RD, et al. *J. Clin. Invest.* 2008; 118 (12), 4025.

[22] Couse JF, et al. *Endocrinology.* 1997; 138, 4613.

[23] Dai SY, et al. *Biochemistry.* 2009; 48, 9668.

[24] Dammann CEL, et al. *Endocrinology.* 2000; 141, 2923.

[25] Davis PB, et al. *Am. J. Respir. Crit. Care Med.* 1996; 154, 1229.

[26] Delaunay F, et al. *Mol. Pharmacol.* 2000; 58, 584.

[27] Douin-Echinard V, et al. J. *Leukoc. Biol.* 2011; 90, 145.

[28] Evans RM. *Science.* 1988; 240, 889.

[29] Fahy JV. *Nat. Rev. Immunol.* 2015; 15, 57.

[30] Fajac I, et al. *Pharmacol. Therapeut.* 2017; 70, 205.

[31] Farrell PM. *J. Cyst. Fibros.* 2008; 7, 450.

[32] Fleisher B, et al. *Obstet. Gynecol.* 1985; 66, 327.

[33] Genuneit J. *Am. J. Respir. Crit. Care Med.* 2014; 190, 588.

[34] Gorski J, et al. *Environ. Health Perspect.* 1995; 103, 69.

[35] Greco TL, et al. *Endocrinology.* 1991; 129, 1326.

[36] Greco TL, et al. *Endocrinology.* 1992; 130, 421.

[37] Green S, et al. *J. Steroid Biochem.* 1986; 24, 77.

[38] Green S, et al. *Nature.* 1986; 320, 134.

[39] Grese TA, et al. *Proc. Natl. Acad. Sci. USA* 1997; 94, 14105.

[40] Hamilton KJ, et al. *Reprod. Biol.* 2014; 14, 3.

[41] Harris HA. *Mol. Endocrinol.* 2007; 21, 1.

[42] Ivanova MM. *Am. J. Respir. Cell Mol. Biol.* 2010; 42, 320.

[43] Kawai H, et al. *Clin. Cancer Res.* 2005; 11, 5084.

[44] Kent AL, et al. *Pediatrics.* 2012; 129, 124.

[45] Khoury MJ, et al. *Am. J. Obstet. Gynecol.* 1985; 151, 777.

[46] Kim S, et al. *Annals of Epidemiology.* 2003; 13, 666.

[47] Kitajima Y, et al. *Endocrinology.* 2016; 229, 267.

[48] Knowles MR, et al. *J. Clin. Invest.* 2002; 109, 571.

[49] Koehler KF, et al. *Endocrinology. Rev.* 2005; 26, 465.

[50] Kuiper GGJM, et al. *Endocrinology.* 1997; 138, 863.

[51] Kuiper GGJM, et al. *Proc. Natl. Acad. Sci. USA.* 1996; 93, 5925.

[52] Kumar R, et al. *Steroids.* 1999; 64, 310.

[53] Kumar V, et al. *Cell.* 1987; 24, 941.

[54] Kynyk JA, et al. *Curr. Opin. Pulm. Med.* 2011; 17, 6.

[55] Jensen EV, et al. *Gynecol. Invest.* 1972; 3, 108.
[56] Lamprecht B, et al. *Chest.* 2011; 139, 752.
[57] Laviolette L, et al. *Can. Respir. J.* 2007; 14, 93.
[58] Lemmen JG, et al. *Mech. Develop.* 1999; 81, 163.
[59] Lingappan K, et al. *Am. J. Physiol. Lung Cell Mol. Physiol.* 2016; 311, L481.
[60] Lonni S, et al. *Ann. Am. Thorac. Soc.* 2015; 12, 1764.
[61] Lubahn DB, et al. *Proc. Natl. Acad. Sci. USA* 1993; 90, 11162.
[62] MacGregor JI, et al. *Pharmacol. Rev.* 1998; 50, 151.
[63] Marandi Y, et al. *Arch. Med. Sci.* 2013; 9, 521.
[64] Massaro D, et al. *Am. J. Physiol. Lung Cell Mol. Physiol.* 2006; 290, L866.
[65] Massaro D, et al. *Am. J. Physiol. Lung Cell Mol. Physiol.* 2004; 278, L1154.
[66] Massaro D, et al. *Am. J. Physiol. Lung Cell Mol. Physiol.* 2007; 293, L222.
[67] Massaro GD. *Am. J. Physiol.* 1996; 270, L110.
[68] Matsui H, et al. *Cell.* 1998; 95, 1005.
[69] Matthews J, et al. *Mol. Interv.* 2003; 3, 281.
[70] McCurnin DC, et al. *Am. J. Respir. Crit. Care Med.* 2009; 179, 492.
[71] McShane PJ, et al. *Am. J. Respir. Crit. Care Med.* 2013; 188, 647.
[72] Merkus PJ, et al. *Pediatr. Pulmonol.* 1996; 21, 383.
[73] Métivier R, et al. *Adv. Exp. Med. Biol.* 2008; 617, 129.
[74] Moorman JE, et al. *MMWR Suppl.* 2011; (60 Suppl), 84.
[75] Morani A, et al. J. Intern. Med. 2008; 264, 128.
[76] Morrissey BM, et al. *Clin. Chest Med.* 2004; 25, 361.
[77] Mosselman S, et al. *FEBS Letters.* 1996, 392, 49.
[78] Nilsson S, et al. *Physiol. Rev.* 2001; 81, 1535.
[79] O'Driscoll DN, et al. *Acta Pædiatrica.* 2018; 107, 1494.
[80] Paech K, et al. *Science.* 1997; 277, 1508.
[81] Patrone C, et al. *Mol. Cell. Biol.* 2003; 23, 8542.
[82] Pike AC, et al. *Steroid Biochem. Mol. Biol.* 2000; 74, 261.
[83] Pike AC, et al. *Biochem. Soc. Trans.* 2000; 28, 396.
[84] Pinkerton KE, et al. *Am. J. Respir. Crit. Care Med.* 2015; 192, 11.

[85] Planey SL, et al. *J. Recept. Signal Transduct. Res.* 2014; 34, 1.
[86] Prescott E, et al. *Eur. Respir. J.* 1997; 10, 822.
[87] Raghavan D, et al. *Respirology.* 2016; 21, 449.
[88] Rastinejad, et al. *Trends Biochem.* Sci. 2015; 40, 16.
[89] Revankar CM, et al. *Science.* 2005; 307, 1625.
[90] Ringbaek T, et al. *Eur. Respir. J.* 2005; 25, 891.
[91] Riordan JR, et al. *Science.* 1989; 245, 1066.
[92] Rowe SM, et al. *N. Engl. J. Med.* 2005; 352, 1992.
[93] Saint-Criq V, et al. *Steroids.* 2014; 81, 4.
[94] Salvi SS, et al. *Lancet.* 2009; 374, 733.
[95] Schatz M, et al. *J. Allergy Clin. Immunol.* 1988; 81, 509.
[96] Seaborn T, et al. *Trends Endocrinol. Metab.* 2010; 21, 729.
[97] Sin DD, et al. *Proc. Am. Thorac. Soc.* 2007; 4, 671.
[98] Skowron KJ, et al. *Mol. Cell Endocrinol.* 2019; 493, 110471.
[99] Sørheim IC, et al. *Thorax.* 2010; 65, 480.
[100] Stevenson DK, et al. *Arch. Dis. Child Fetal Neonatal.* 2000; 83, F182.
[101] Sunyer J, et al. *Am. J. Respir. Crit. Care Med.* 1998; 158, 851.
[102] Sweezey NB, et al. *Pediatr. Pulmonol.* 2014; 49, 309.
[103] Takeyama J, et al. *J. Clin. Endocrinol. Metab.* 2001; 86, 2258.
[104] Tam A, et al. *Am. J. Respir. Crit. Care Med.* 2016; 193 (8), 825.
[105] Tam A, et al. *BMC Women's Health.* 2011; 11, 24.
[106] Taylor AH, et al. *J. Mol. Endocrinol.* 2000; 24, 145.
[107] Tenbaum S, et al. *Int. J. Biochem. Biol.* 1997; 29, 1325.
[108] Tora L, et al. *Cell.* 1989; 59, 477.
[109] Tóth P, et al. *Orv. Hetil.* 2003; 144, 1719.
[110] Townsel CD, et al. *Front. Pediatr.* 2017; 5, 1.
[111] Troisi RJ, et al. *Am. J. Respir. Crit. Care Med.* 1995; 152, 1183.
[112] Trotter A, et al. *Ann. Med.* 2000; 32, 608.
[113] Trotter A, et al. *Early Human Dev.* 2009; 85, 353.
[114] Trotter A, et al. *Int. J. Pediatr.* 2009; 2009, 170491.
[115] Trotter A, et al. *Paediatr. Drugs.* 2001; 3, 629.
[116] Trotter A, et al. *Arch. Dis. Child.* Fetal Neonatal Ed. 2007; 92, F94.
[117] Tulchinsky D, et al. *Am. J. Obstet. Gynecol.* 1972; 112, 1095.

[118] Tzukerman MT, et al. *Mol. Endocrinol.* 1994; 8, 21.
[119] van Hoorn WP. *J. Med. Chem.* 2002; 45, 584.
[120] Vaskivuo TE, et al. *J. Clin. Endocrinol. Metab.* 2005; 90, 3752.
[121] Vidaillac C, et al. *Breathe.* 2018; 14, 109.
[122] Wang Y, et al. *Proc. Natl. Acad. Sci. USA.* 2006; 103, 9908.
[123] Webster NJ, et al. *Cell.* 1988; 15, 199.
[124] Weihua Z, et al. *Proc. Natl. Acad. Sci. USA.* 2001; 98, 6330.
[125] Weihua Z, et al. *Proc. Natl. Acad. Sci. USA.* 2000; 97, 5936.
[126] Wu W, et al. *J. Allergy Clin. Immunol.* 2014; 133, 1280.
[127] Ying C, et al. *Mol. Reprod. Dev.* 2000; 55, 83.
[128] Yu X, et al. *Am. J. Physiol. Endocrinol. Metab.* 2011; 301, E882.
[129] Yung JA, et al. *Ann. Allergy Asthma Immunol.* 2018; 120, 488.
[130] Zein JG, et al. *Curr. Allergy Asthma Rep.* 2015; 15, 28.
[131] Zeitlin PL. *J. Clin. Invest.* 2008; 118, 3841.
[132] Zisk JL, et al. *Am. J. Perinatol.* 2011; 28, 241.

In: Estrogen Receptors
Editors: George G Chen et al.
ISBN: 978-1-53618-228-6

Chapter 7

ESTROGEN RECEPTOR ALPHA 36 IN CANCERS

Lingbin Xue[1,4], Alexander C. Vlantis[1,4], MD, Jason Y. K. Chan[1,4], MD, Qian Cao[1,5], Xianhai Zeng[3,4], Fang Ma[3,4], C. A. van Hasselt[1,4], MD, George G. Chen[1,2,4,*], PhD and Michael C. F. Tong[1,2,4,†], MD

[1]Department of Otorhinolaryngology, Head and Neck Surgery, The Chinese University of Hong Kong (CUHK), Prince of Wales Hospital, Shatin, NT, Hong Kong, China
[2]CUHK Shenzhen Research Institute (SZRI), Shenzhen, Guangdong, China
[3]Shenzhen Key Laboratory of ENT, Institute of ENT & Longgang ENT Hospital, Shenzhen, Guangdong, China,
[4]The Chinese University of Hong Kong – Shenzhen Ear, Nose and Throat Joint Research Centre, Shenzhen, Guangdong, China,
[5]Department of Pediatrics, The First Affiliated Hospital, Nanjing Medical University, Nanjing, Jiangsu, China

[*] Corresponding Author's Email: gchen@cuhk.edu.hk.
[†] Corresponding Author's Email: mtong@ent.cuhk.edu.hk.

ABSTRACT

Estrogen receptor α 36 (ER-α36) is derived from the traditional ER-α66 and plays an important role in cellular functions. There are no transcriptional activation function 1 (AF-1) and 2 (AF-2) domains at the 5' terminal and 3' terminal of ER-α36, but the DNA-binding-domain, the nuclear localization signal domain and the ligand-binding domain (LBD) of ER-α66 are preserved. ER-α66 is located in the nucleus, while ER-α36 is mainly found on the cell membrane and in the cytoplasm. ER-α36 belongs to the family of membrane estrogen receptors (ERs) and has strong ligand-dependent activity, which can rapidly react to endogenous and exogenous estrogens. In this chapter, the structure, biological functions, and therapeutic potentials of ER-α36 are reviewed in related diseases. We aim to provide a reference value for the future research of ER-α36.

Keywords: estrogen receptor, ER-α36, non-genomic signaling, cancer

INTRODUCTION

Estrogen plays important roles not only in physiological conditions such as body growth, reproduction, and organ function maintenance, but also in pathological conditions including osteoporosis, atherosclerosis, Alzheimer's disease and in many types of cancers [1]. Estrogen, mainly 17-β estradiol (E2), is synthesized by aromatase-mediated catalyzation of androgen, mainly in the inner cells and granulosa cells of ovarian follicles [2]. E2 can regulate the proliferation, differentiation, and homeostasis of eukaryotic cells. It is well known that E2 and/or its receptors are associated with a high risk of certain cancers in primary estrogen targeted organs such as breast and the endometrium [3, 4]. Interestingly, on the other hand, they may also exhibit a protective role in secondary estrogen targeted organs such as the liver and thyroid [5]. Accumulating evidence has indicated that the different results may be caused by the diverse isoforms of ERs [5-7].

Estrogen exerts its functions through ERs [1]. Two forms of ERs, ER-α and ER-β were successively cloned in 1986 and 1996 respectively [8, 9]. The wild-type ER-α (66kDa), also called ER-α66, is mainly distributed in the nucleus and mediates the genomic estrogen signaling pathway [7]. Later, a short-form ER-α46 (46kDa) was detected in the nucleus and cytosol and was confirmed to have high membrane affinity and the ability to mediate the endothelial nitric oxide synthase (eNOS) pathway in endothelial cells [10, 11]. In 2005, Wang *et al.* identified another ER variant (35.7kDa) and named it ER-α36, which was transcribed from a promoter located in the first intron of the ER-α66 gene [12]. ER-α36 lacks both transcriptional activation domains (AF-1 and AF-2) but retains the DBD, dimerization, and partial LDB [12].

Unlike the traditional nuclear receptors ER-α66 and ER-β, ER-α36 lacks intrinsic transcriptional activity and mainly mediates the non-genomic estrogen signaling pathway [13]. Both estrogen and anti-estrogen agents may activate ER-α36. Activated ER-α36 can promote breast cancer cell proliferation through the caveolin-1-mediated mitogen-activated protein kinase (MAPK) signaling pathway [13, 14]. Signal pathways activated by estrogen and anti-estrogen through ER-α36 may help us to identify why some human cancers resist anti-estrogen therapies. In this chapter we review the recent advances in ER-α36 biology and function in estrogen-dependent and independent tumors.

THE STRUCTURE OF ER-A36 AND OTHER ER-A ISOFORMS

The traditional wild-type ER-α protein which consists of 595 amino acids with a molecular weight of 66-kDa, is described as the ESR1 gene (6q25.1) and called ER-α66 [6]. In 2005, Wang *et al.* found a full-length cDNA clone gene from human mammary epithelial caveolin-1 cells, which fully matched the DNA sequence of the ER-α66 ligand-dependent region gene transcription. The total length of this cDNA is 5.4kb, which can be transcribed into 310 amino acid open reading frames, encoding a protein with a molecular weight of about 35.7-kDa, namely ER-α36 [12]. Sequence analysis has indicated that the ER-α36 isoform originates from exons 2 to 6 of the ESR1 gene, and that the last 27 amino acids are encoded by exon 9 (Figure 1) [12, 15]. This new estrogen receptor subtype has both similarities to and a uniqueness from the traditional estrogen receptor. The traditional

ER-α and ER-β are members of the nuclear receptor family, having the common structural characteristics of the family. Its primary structure can be divided into five functional areas from the N-end to the C-end: A/B, C, D, E, and F (Figure 1). The A/B domain contains an AF-1 that is not ligand-dependent. The C region is a DNA binding domain (DBD). The E region is a LBD that contains a ligand-dependent transcriptional AF-2. The D region is a variable hinge on the domain that connects DBD and LBD. The biological function of the F region is still unclear [16].

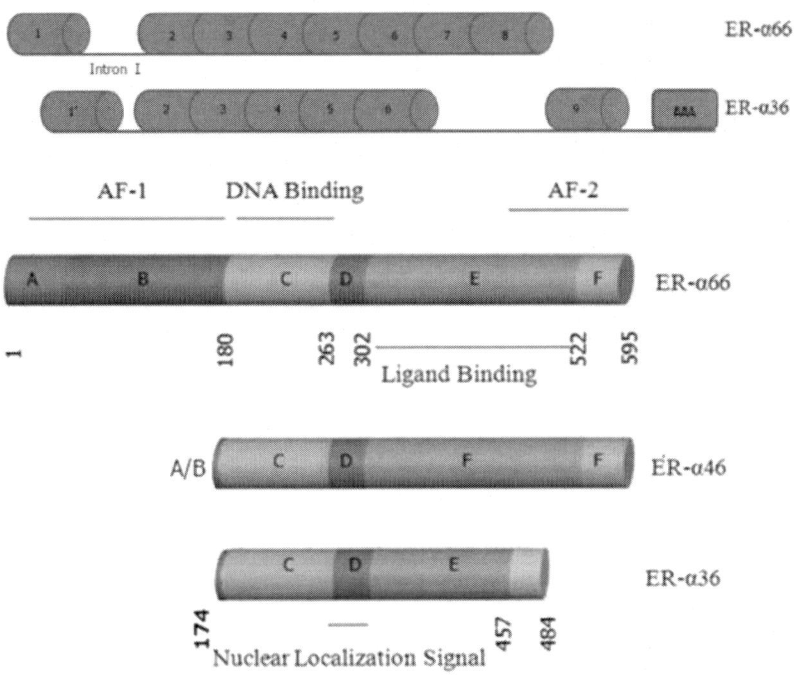

Figure 1. The protein structure of ERs. The protein structure of ERs includes the A/B domain, transcriptional AF-1, C domain contains DBD, D domain that represents nuclear localization signal (NLS), E domain that corresponds to ligand binding domain (LBD) and transcriptional AF-2. ER-α36 lacks AF-1 and AF-2. The last 27 amino acids of ER-α36 are indicated by a grey box. Nucleotides in ESR1 gene and amino acids in ER-α66 are numbered. ER-α36 contains amino acids corresponding to residues numbered 174–457 in ER-66. Additional 27 amino acids are encoded by exon 9.

ER-α36 lacks the transcriptional activation region (AF-1 and AF-2) of ER-α66 but retains DBD, partial dimer, and LBD (Figure 1). Although ERα36 has no transmembrane domain, it has a growth factor receptor-like function binding to a variety of estrogens. The LBD of ER-α36 can be modified by post-transcriptional palmitoylation and contains three potential sites for myristoylation [17, 18]. Therefore, ER-α36 can be coupled to some membrane proteins and anchored on the plasma membrane to mediate the membrane initiation estrogen signal [19] or non-genomic estrogen signal pathways [20].

In 2000, a 46-kDa ER-α isoform was cloned by Flouriot *et al.* [21]. This demonstrated that the 46-kDa subtype lacks the first 173 amino acids (A/B or AF-1 domain) and is a selection of the ER-α gene obtained by skipping exon 1 (Figure 1). ER-α46 exists in the plasma membrane, cytoplasm, and nucleus. It can mediate rapid estrogen signaling including the activation of Src/phosphatidylinositol 3 kinase (PI3K)/serine threonine kinase (AKT) signals and stimulation of NO synthesis [22]. All these data suggest that the ER-α46 isoform is related to the rapid estrogen signaling pathway [11].

The Biology of ER-α36

The traditional estrogen receptor ER-α66 is a nuclear receptor, mainly distributed in the nucleus, which belongs to the nuclear receptor family. The transcription factors existing in the nucleus are activated by ligands to execute their biological effects. The fact that ER-α36 has no intrinsic transcriptional regulation suggests that ER-α36 may mediate rapid estrogen signaling that can work without entering the nucleus. Sub-cellular fractionation assay of ER-α36 transfected HEK293 cells has shown that a high proportion of ER-α36 (~ 50%) is located at the plasma membrane with a low proportion in the cytoplasm (~ 40%) and the nucleus (~ 10%) [12]. Immunohistochemical analysis of specimens from breast cancer patients has also shown that ER-α36 is mainly expressed outside the nucleus [23]. According to experimental results of the osteoblast membrane, unlike the classical estrogen receptor ER-α66, ER-α36 is a membrane receptor [24], which is widely distributed on or near the cell membrane and receives signal transduction from extracellular and intracellular sources. ER-α36, as a membrane receptor, has different biological effects from ER-α66 in physiological and pathological processes.

There are some interactions among ER-α isoforms. Wang *et al.* have demonstrated that ER-α36 could inhibit the genomic signal transduction pathway of ER-α66 and ER-β and that ER-α66 could suppress the activity of AP1 promoter of ER-α36, which could be relieved by ER-α36 and ER-α46 [7]. Besides, ER-α46 can inhibit the transcriptional activity of the AF-1 domain of ER-α66 [7]. Therefore, the levels of ER-α66, 46, and 36 will determine whether the cell chooses the genome or non-genome signal transduction pathway of estrogen. Both E2β and BSA-conjugated E2β (an estradiol that binds bovine serum albumin and has cell membrane impermeability) can induce rapid estrogen signaling such as activation of the MAPK/extracellular signal-regulated kinase (ERK) pathway, PI3K/AKT pathway, and stimulation of cell proliferation in HEK293 cells transfected with ER-α36, without being affected by tamoxifen (TAM), 4-hydroxy AKT, and ICI 182,780 [12], supporting the fact that the estrogen-mediated the non-genomic signaling is via the membrane ER-α36 rather than ER-α66. Other estrogens including estrone (E1), estriol (E3) and estradiol (E4) can also function similarly to E2 to activate ER-α36. It is suggested that ER-α36 may have a wider ligand binding spectrum than ER-α66 [12]. ER-α36 was reported to mediate estrogen activation of the protein kinase C (PKC)δ/ERK signaling pathway in endometrial cancer Hec1A cells [25]. Overall, ER-α36 is a new type of ER that mediates rapid and non-genomic estrogen signaling.

Pathways of Er-a36

ER-α36 and MAPK/ERK and PI3K/AKT Pathways

The MAPK/ERK pathway plays an important role in the occurrence and development of many tumors, such as breast, prostate, and colorectal cancers [26]. Wang *et al.* found that the phosphorylated ERK level was significantly increased in HEK293 cells transfected with ER-α36 compared with the control group when cells were treated with E2β-BSA [19], indicating that ER-α36 can activate the MAPK/ERK signal pathway. MAPK can be activated by the phosphorylation of specific threonine and tyrosine residues of MAPK kinases. MAPK kinases themselves are activated by their upstream protein kinase Raf that is regulated by the Ras protein family. Interestingly, the low concentration of estrogen accelerates tumor progression on triple-negative breast cancer cells through activation of the oncogene Src, and by phosphorylating the tyrosine kinase site Src-Y416. However, under the high concentration of estrogen, Src-Y527 is phosphorylated to inhibit tumor cell growth [27]. Lin *et al.* found that on treatment with testosterone, E2, or AKT could activate the PI3K/AKT signal pathway by phosphorylating AKT in endometrial cancer Hec1A cells, but that this signal pathway could not be activated after ER-α36 was knocked out [25], indicating that the presence of ER-α36 is necessary for these stimulating agents. Tu *et al.* showed that ER-α36 can work with epidermal growth factor (EGF) to trigger the phosphorylation of downstream ERK1/2 [28]. It can thus be seen that ER-α36 is closely related to MAPK/ERK and PI3K/AKT pathways and plays an important role in the proliferation and drug resistance of cancer cells [7, 29].

ER-α36 and EGFR/HER2 Pathway

Both epithelial growth factor receptor (EGFR) and HER2 are important targets in cancer therapy and are involved in the regulation of various signaling pathways. Zhang *et al.* found that there was a positive feedback loop between ER-α36 and EGFR in breast cancer [29]. They have demonstrated that the expression of ER-α36 plays an important role in maintaining EGFR protein stability and that the EGFR pathway can activate the transcription of ER-α36 through the binding of AP-1 in the promoter region of ER-α36 gene [29]. The positive association between both molecules have been further confirmed by others [30, 32]. Tu *et al.* found that the expression of ER-α36 was positively correlated with the level of EGFR in endometrial cancer [28]. A similar positive feedback loop appears between ER-α36 and HER2 and has been reported in AKT-resistance breast cancer, and so it is concluded that ER-α66 negative breast cancer can promote cell proliferation and resist AKT therapy through the synergistic effect of ER-α36-EGFR/HER2 pathway [14]. The detailed signaling pathways between ER-α36-EGFR/HER2 are currently unknown and whether the synergistic effect of ER-α36 and EGFR/HER2 exists in other cancers deserves to be explored.

ER-α36 and Ca^{2+} Pathway

Ca^{2+} and cAMP are widely present in cells as secondary messengers involved in the regulation of cell activities and functions. Kang *et al.* measured the Ca^{2+} flow in E2-treated human breast cancer cells SK-BR-3 and found that the Ca^{2+} concentration was significantly increased in cells containing ER-α36. However, when ER-α36 was inhibited, E2 was unable to stimulate Ca^{2+} [13]. Therefore, it is concluded that ER-α36 can positively regulate intracellular Ca^{2+} concentrations [13]. The clinical significance of ER-α36-mediated up-regulation of the Ca^{2+} concentration is worth studying.

ER-A36 IN CANCERS

Breast Cancer

The breast is one of the major estrogen-targeting organs and breast cancer is one of the most common female tumors [30]. Due to appropriate anti-estrogen therapies, about 80% of ER-positive breast cancer patients can survive for 5 years or more [31]. Currently, anti-estrogen therapy drugs targeting ER-α include selective estrogen receptor modulators (SERMs) such as AKT (TAM) and pure anti-estrogens such as ICI 182780 [32]. TAM is able to block estrogen signaling by competing with ER-α66 to combine with estrogen as a complex, and thus inhibits estrogen function and the proliferation of estrogen-dependent breast cancer cells [33]. Despite the significant anti-estrogen activity of anti-estrogen therapy, 20% to 30% of patients will develop drug resistance while using TAM [34]. Studies have indicated that anti-estrogen resistance is often associated with G protein-coupled estrogen receptor 30 (GPR30) and/or ER especially ER-α36 [35].

The expression of ER-α36 can be detected in both ER-positive and -negative breast cancers. A retrospective study of 896 cases of breast cancer revealed that about 40% of all the cases were positive for ER-α36, specifically, about 40% of cases co-expressed ER-α66 and ER-α36 and about another 40% of cases of ER-α66-negative breast cancer were also positive for ER-α36 [36]. Interestingly, patients with ER-α66–positive tumors that also express high levels of ER-α36 were less likely to benefit from TAM treatment [36], suggesting that the presence of ER-α36 has a negative impact on the anti-estrogen treatment.

Although most breast cancers are estrogen-dependent tumors, anti-estrogen therapy can reduce the growth of the tumors by reducing the levels of estrogen or by inhibiting the effects of estrogen [37]. A significant number of breast cancer patients either initially do not respond to or subsequently become non-responsive to anti-estrogen therapies, the former being termed primary anti-estrogen resistance and the latter secondary anti-estrogen resistance. Wang *et al.* reported that breast cancer cells with high expressions of ER-α36 were highly resistant to TAM, but that the sensitivity of TAM could partially be recovered with ER-α36 silence [12], suggesting that ER-α36 contributes, at least in part, to TAM resistance.

The mechanism leading to anti-estrogen resistance is complicated. However, to date, ER-α36 has been considered as an important factor causing primary or secondary anti-estrogen resistance via the following five pathways. Firstly, while TAM inhibits the proliferation and growth of cancer cells, it also stimulates the MAPK/ERK activity and the increasing MAPK/ERK activity gradually enables cancer cells to develop resistance to TMA. ER-α36 has been reported to contribute to the resistance by facilitating MAPK/ERK activity [38]. Secondly, ER-α36 can take part in the process of TAM-induced myeloid tumor oncogene c-Myc expression, which may also promote the development of resistance [39.40]. Thirdly, regardless of being ER-positive or –negative breast cancer cells, the high expression of

ER-α36 during TAM treatment leads to the activation of the ER-α36-EGFR/HER2 positive regulatory loops [7], which can change the breast cancer cells from estrogen-dependent to growth factor-dependent and escape the effectiveness of TAM. Fourthly, ER-α36 can phosphorylate EGFR and Src via physically interacting with the EGFR/Src/Shc complex, and EGFR can also promote the transcription of ER-α36 through the AP1 site on the ER-α36 promoter, creating a positive circulation loop to keep the expression of EGFR at a stable level. This ER-α36-EGFR/HER2 positive regulatory loop stimulates a series of downstream pro-growth proteins such as STAT3/5 and Cyclin D1 [28]. Finally, ER-α36 can cause breast cancer cells to develop resistance to anti-estrogen treatments by stimulating cancer stem cells that are a small subset of cancer cells but which significantly contribute to cancer cell differentiation, self-renew, and selective resistance to various drugs [41]. Deng *et al.* found that the expression of ER-α36 was increased during the development of MCF7 and T47D cancer stem cells and that these cells resisted TAM treatment [42]. They also showed that after treating ER-α36-positive cancer stem cells with TAM, the number of CD44+/CD24- cells increased. However, the block of ER-α36 by its siRNA could significantly reduce these CD44+/CD24- cells leading to cancer cells becoming less resistant to TAM treatment [42]. Wang *et al.* also proved that AKT activates ER-α36 to enhance the stemness and metastasis of breast cancer cells by TAM [43]. Therefore, it is believed that ER-α36 can activate at least these five different pathways to contribute to anti-estrogen resistance.

Endometrial Cancer

Endometrial cancer is a common gynecological tumor. It is generally considered that endometrial cancer can be estrogen-dependent or non-estrogen-dependent. Among them, estrogen dependence is related to excessive estrogen stimulation and estrogen receptor expression.

Tu *et al.* found that ER-α36 and EGFR were co-localized on the cell membrane of endometrial cancer Hec 1A cells by immunofluorescence and

Western blotting [28]. The study also found that ER-α36 positively associates with high-stage and high-grade endometrial cancer. Furthermore, this research reported that ER-α36 and EGFR expression had higher levels of phosphorylation of extracellular regulated kinase (ERK) than samples with negative ER-α36 and EGFR expression, which indicates that ER-α36 regulates the activation of ERK by epithelial growth factor (EGF) in Hec 1A cells. The expression of ER-α36 protein has been measured by immunohistochemistry in 73 cases of endometrial cancer, 20 cases of normal endometrial tissues, and 9 cases of endometrial atypical hyperplasia [4]. The expression of ER-α36 in endometrial carcinoma tissues was significantly lower than in normal endometrial and endometrial tissues with atypical hyperplasia ($p < 0.01$).

Long-term and non-antagonistic estrogen stimulation is the main mechanism promoting endometrial cancer. These results have shown that estrogen stimulates endometrial cell proliferation by binding with the plasma membrane ER, particularly ER-α36, playing an important role in the development and growth of endometrial cancer [44]. E2 can bind to the plasma membrane ER-α36 to initiate its downstream pathways including PKCδ/ERK- cyclin D1/cdk4, MAPK/ERK and PI3K/AKT signaling, stimulating the proliferation and growth of endometrial cancer cells [44]. The disease-free survival rate of ER-α36-positive patients is lower than ER-α36-negative patients ($P < 0.01$). However, the expression of ER-α36 is not related to age, surgical stage, tissue differentiation, myometrial infiltration, lymph node metastasis (LNM), nor pathological type in endometrial cancer ($P > 0.05$) [38]. In summary, it can be inferred that ER-α36 can be a therapeutic target in estrogen-dependent endometrial cancer. ER-α36 may be a useful biomarker for the diagnosis, prognosis, and treatment of endometrial cancer.

Cervical Cancer

Cervical cancer is another common gynecological tumor, and is caused by infection with human papillomaviruses. It has been confirmed that high estrogen levels may be also an inducement of cervical cancer by clinical observational studies and animal studies [45.46]. Sun *et al.* reported the expression of ER-α36 was found in both cervical squamous cell carcinoma human tissue and two human papillomavirus-infected human cervical cancer cell lines (CaSki and HeLa) [47]. This research also showed that the down-regulation of ER-α36 significantly inhibited cell invasion, migration, and proliferation. On the other hand, the upregulation of ER-α36 plays the opposite role. ER-α36 mediates estrogen-stimulated MAPK/ERK activation.

Zhang *et al.* reported that the expression level of ER-α36 in HeLa cells decreased significantly after the treatment of Compound Kushen injection (a traditional Chinese medicine), especially when combined with cisplatin for 48 hours [48]. It can be concluded that ER-α36 is closely related to the proliferation of HeLa cells, and that inhibition of ER-α36 expression may be of great importance in the treatment of cervical cancer.

Papillary Thyroid Cancer

Thyroid cancer is of less concern because of its relatively slow progression and better prognosis. However, in the past 20 years, there has not been a breakthrough in the non-surgical treatment of thyroid cancer metastases, nor has its recurrence rates and mortality been significantly improved. In recent years, this has been a research hotspot - to find the diagnosis index and treatment target of thyroid cancer metastases [49, 50].

Papillary thyroid carcinoma (PTC) accounts for 80% of thyroid malignancy [51]. Worldwide, the incidence of PTC in women is about three to four times that of men [52]. Studies have shown that its occurrence and development is related to estrogen and/or its receptor levels. *In vitro* experiments have confirmed that E2 can significantly up-regulate the expression of ER in PTC cells, and stimulate tumor cell proliferation, invasion, and metastasis [53]. The thyroid is not the target organ of estrogen in the traditional sense, but more and more evidence has shown that estrogen and ER play an important role in the occurrence and development of thyroid diseases.

It has been shown that estrogen may be involved in its progression, similar to the situation in the breast, ovarian and endometrial cancer. Dai *et al.* found that the expression of ER-α36, glucose-regulated protein (GRP78), and glucose-regulated protein (GRP94) proteins was up-regulated in the primary PTC tissues and was involved in the proliferation, invasion, and migration of PTC-derived BCPAP cells [54]. It was found that high ER-α36 expression combined with high GRP78/GRP94 levels was more associated with extra thyroid extension (ETE), LNM, distant metastasis (DM), and high TNM staging than the combination of high ER-α36 with high GRP78 or high GRP94 expression. ($p = 0.028$ for ETE, $p = 0.002$ for DM, $p \leq 0.001$ for LNM and TNM high staging) or only ER-α36 expression (for ETE, LNM, DM, and High TNM staging, $p < 0.001$).

To determine the expression of ER-α36 in primary PTC, 68 PTC specimens were examined by immunohistochemistry [55]. That study confirmed that the expression of ER-α36 was significantly higher than ER-α66 (67.6% vs 20.6%, $P < 0.01$), especially in people over 45 years old (95.2% vs 55.3%, $P < 0.01$). This finding is in line with the concept that age is an important factor influencing the metastasis, recurrence, and death of thyroid cancer [56]. The 2018 National Comprehensive Cancer Network (NCCN) guidelines for thyroid cancer staging standards for PTC patients indicate that staging in patients over 45 years old is significantly worse than in patients under 45 [57].

Therefore, ER-α36 may provide new clues to explain the occurrence and development of thyroid cancer, especially the high incidence in females, and may become a new diagnostic indicator and target for thyroid cancer treatments.

Hepatocellular Cancer

Hepatocellular carcinoma (HCC) is a disease with many triggering factors and genetic variations. The liver is also a hormone-sensitive organ, although it is generally considered to be an accessory organ of the digestive system. ERs are expressed in human HCC [58].

HCC is a cancer with a worldwide gender disproportion as the incidence, morbidity and mortality of HCC is much higher in males than in females. Therefore, sex hormones have been suggested to be involved in regulating the malignant transformation of hepatocytes [5]. The epidemiological evidence has also been supported by clinical observation or laboratory data that postmenopausal females with high levels of circulating estrogen have a better prognosis with HCC [59, 60]. Chen *et al.* have shown that ER-α can interact directly with the β-catenin promoter region following E2 treatment [61]. This result proved that estrogen and ER-α blocked the metastatic function of HA22T cells by modulating GSK3β and βTrCP expression [61]. Therefore, to a certain extent, estrogen and/or ER-α can inhibit the growth and metastasis of HCC.

However, ER-α36 has shown different influences in HCC. Zhang *et al.* reported that the expression pattern of ER-α36 in primary HCC is higher than in secondary HCC [62], suggesting that ER-α36 may be a biomarker to distinguish primary and secondary HCC. Cocciadiferro *et al.* found that the expression of neurofibromatosis type 2 (NF2) gene was elevated in HCC tissues and HepG2 cells but lower in non tumoral liver cells [63]. NF2 expression was also positively correlated with ER-α36 but inversely related to wild-type ER-α66, indicating that the function of ER-α36 is different from that of ER-α66, with the former being pro-HCC and the latter being anti-HCC [63]. Increased ER-α36 in liver tumors was also reported by Lau *et al.* [64]. Furthermore, Teng *et al.* found that dehydroepiandrosterone could increase ER-α36 mRNA and protein levels to up-regulate the oncomiR miR-21 in the liver [65]. ER-α36 was also confirmed as the main ER in HCC cells and served as a biomarker of responsiveness to Epigallocatechin-3-gallate (EGCG) inhibition [66]. Shi *et al.* proved that ER-α36 negatively regulated the transcriptional activity of ER-α66 and ER-β [67]. These results suggest that ER-α36 is highly expressed relative to ER-α66, and that it may have diagnostic and therapeutic values in HCC.

Laryngeal Cancer

Laryngeal cancer is a common head and neck tumor. Due to the lack of specific cell markers, it is still difficult to diagnose and evaluate the prognosis of laryngeal cancer [68]. Epidemiological studies have shown that sex hormones may be involved in the occurrence of laryngeal cancer due to the great difference in gender sensitivity, with the ratio of men to women being 11:1 [69]. The larynx is a secondary sex organ, which undergoes trophic changes in response to hormone changes in adolescence and changes in its morphological structure in adulthood. Laryngeal cancer is still a controversial topic regarding hormone therapy.

Schwartz *et al.* [70] studied the laryngeal cancer cell line Hep2 *in vitro*, and found that ER-α36 can rapidly activate the PKC cell pathway and phospholipase D (PLD) activity by binding with E2, and promote the proliferation of tumor cells. In addition, the activation of ER-α36 by E2 can up-regulate the expression of metastatic factors. The above results suggest that ER-α36 might be a new target for the treatment of laryngeal cancer.

Gastric Cancer

Gastric cancer (GC) is one of the most common cancers worldwide [71]. Epidemiological evidence shows that the incidence of GC in men is higher than that in women [72]. The occurrence of GC in males is 10-15 years earlier than in females; however, after menopause, it is equally prevalent in both males and females [72]. Furthermore, ovarian or exogenous estrogen may suppress GC and estrogen exposure can reduce the risk of GC [73, 74]. Therefore, estrogen may play a protective role in the development of GC.

However, unlike the protective role of the traditional ER-α66, ER-α36 appears to be an oncogenic molecule in GC. The ER-α36 expression can be detected in gastric cell lines and primary cancer tissues [75, 76]. Wang *et al.* found that the expression of ER-α36 in GC cell lines is higher than that of ER-α66 [76]. ER-α36 is highly expressed in cancer tissues, while ER-α66 is highly expressed in normal tissues, suggesting that ER-α36 may be oncogenic whereas ER-α66 may be suppressive against GC. It was reported that the invasion ability of GC cells was positively correlated with the expression of ER-α36. It is likely that ER-α36 and miR-143 jointly participate in up-regulating the invasion of GC cells [77]. Fu *et al.* proved that ER-α36-mediated rapid estrogen signaling positively regulates GRP78 expression, which may be associated with the growth of GC [78]. Based on a 117 case study of GC, Wang *et al.* reported that ER-α36 expression was correlated with gender (male: female ratio 2.88:1, $P = 0.01$), invasion to the serosa ($P = 0.01$) as well as cyclin D1 expression ($P < 0.01$) [79].

Therefore, ER-α36 plays a positive role in the proliferation and growth of GC cells while ER-α66 may inhibit GC. Nevertheless, the role of ER-α36 in GC has not been fully defined, and its clinical significance largely remains to be explored.

Non-Small Cell Lung Cancer

ER-α is highly expressed in non-small cell lung cancer (NSCLC) [80, 81]. Also, Zhang *et al.* [82] analyzed the expression of ER-α36 and ER-α66 in 126 cases of primary NSCLC by Western blotting and immunohistochemistry and found that the expression of ER-α36 was higher than that of ER-α66. The study also found that ER-α36 was highly expressed and positively correlated with LNM in lung adenocarcinoma, but lowly expressed in lung squamous carcinoma. These results indicate that ER-α36 may play a role in the promotion of lung adenocarcinoma.

Colorectal Cancer

Colorectal cancer is one of the most common malignant tumors of the digestive system. Epidemiological studies have shown that men of all ages are more likely to develop colon cancer than women [83, 84]. Hormone replacement therapy can reduce the risk of colorectal cancer by 30% to 40%, and prevent the occurrence of colorectal cancer in postmenopausal women

[85, 86]. These reports suggest that estrogen plays a preventive role in the development/growth of colorectal cancer.

Jiang *et al.* analyzed 35 cases of colorectal cancer tissues and found that the expression of ER-α36 was 25/35(71%) [87]. However, the mRNA level of ER-α36 in colorectal cancer tissues was lower than normal tissues. Furthermore, the expression of ER-α36 mRNA was negatively correlated with tumor stage and LNM. Therefore these results appear to suggest that ER-α36 functions as a tumor suppressor in colorectal cancer, which is different from the oncogenic role of ER-α36 in other cancers.

Glioblastoma

Glioblastoma (GBM) is the most common primary malignant brain tumor. Recent studies have indicated that TAM treatment can sensitize GBM cells to radiotherapy and chemotherapy [88, 89]. However, a significant number of GBM patients, especially the recurrent or long-term TAM-treated patients, showed little response to TAM treatment [90, 91]. It was reported that TAM can induce apoptosis in GBM cells via the PI3k/AKT, JNK, and ERK pathways [92]. However, the molecular mechanisms of TAM resistance remain unknown.

Liu *et al.* reported that ER-α36 was detectable in both GBM cell line U87-MG and U251 [93]. The down-regulation of ER-α36 could inhibit the growth of GBM U87-MG cells, while on the contrary, the up-regulation of ER-α36 could enhance the proliferation [93]. Moreover, ER-α36 was highly expressed (27 of 31, 87.1%) in human GBM tissues, and ER-α36 knockdown increased the sensitivity of U87 cells to TAM and decreased autophagy [94]. These results indicate that ER-α36 may upregulate autophagy and contribute to TAM resistance in GBM cells.

The expression and clinical significance in cancers are summarized (Table 1).

Table 1. The expression of ER-α36 in cancers

Cancer type	No. of Pos./ Tot. (%)	Method	Significance	References
Breast Cancer (BC)	493/1068 (46.2%)	IHC	AKT activates ER-α36 to enhance the stemness and metastasis of BC cells	Wang *et al.* [43]
	366/896 (40.1%)	IHC	ER-α36++/ER-α66+ BC results in a worse curative effect from AKT treatment than does ER-α66+/ER-α36- BC	Shi *et al.* [40]
	7/37 (19%)	IHC/ WB	ER-α36 expression is negatively related to ER-α66	Lee *et al.* [23]
	56/74 (75.7%)	PCR	ER-α36 expression in BC < in matched normal tissues; Down-regulation of ER-α36 mRNA was correlated with local progression, LNM, and advanced cancer stage	Zheng *et al.* *et al.* [95]
	26/30 (86.7%)	IHC	ER-α36 was highly expressed in ER-α66-negative apocrine and adenoid cystic BC	Vranic *et al.* [96]
Endometrial Cancer (EC)	24/45 (53%)	IHC	ER-α36 was positively associated with high-stage/high-grade	Tu *et al.* [28]
	24/73 (32.9%)	IHC	ER-α36 expression in EC < in normal endometrial tissues	Sun *et al.* [4]
Cervical Cancer (CC)	36/40 (90%)	IHC	Down-regulation of ER-α36 significantly inhibited cell invasion, migration, and proliferation in CC	Sun *et al.* [55]
PTC	209/218 (95.9%)	IHC	ER-α36 was highly expressed in PTC; high ER-α36 combined with high GRP78/GRP94 levels was associated with the progress of PTC than was the combination of high ER-α36 with high GRP78 or high GRP94 expression	Dai *et al.* [62]
	46/68 (67.6%)	IHC	The expression of ER-α36 > ER-α66 especially in people over 45 years old	Yang *et al.* [63]
HCC	8/8 (100%)	PCR	ERα36 expression was highly upregulated in HCC compared to normal liver tissue	Miceli *et al.* [66]
	8/8 (100%)	PCR	NF2 expression is positively correlated with ER-α36 but inversely related to wild-type ER-α66, which indicates ER-α36 is highly involved in primary HCC	Cocciadiferro *et al.* [72]
LC	39/39 (100%)	IHC	ER-α36 promotes the proliferation of LC cells	Schwartz *et al.* [79]

Cancer type	No. of Pos./ Tot. (%)	Method	Significance	References
GC	17/22 (77.3%)	PCR	ER-α36 expression in GC > in normal gastric tissues; ER-α36 expression is highly correlated with LNM in GC	Deng *et al.* [84]
	76/86 (88.4%)	IHC	ER-α36 expression in GC > in normal gastric tissues; ER-α36-mediated rapid estrogen signaling positively regulates GRP78 expression, which may be associated with GC	Fu *et al.* [87]
	101/117 (86.2%)	IHC	ER-α36 expression is correlated with gender, invasion to serosa, and Cyclin D1 expression	Wang *et al.* [88]
NSCLC	126/126 (100%)	IHC	The expression of ER-α36 > ER-α66 in NSCLC; ER-α36 is correlated with LNM in adenocarcinoma	Zhang *et al.* [91]
CoC	25/35 (71%)	IHC	ER-α36 expression in CoC < in normal tissues; Inverse association with tumor stage/LNM	Jiang *et al.* [96]
GBM	27/31 (87.1%)	IHC	ER-α36 was highly expressed in GBM; ER - α36 knockdown increased sensitivity of GBM U87 cells to TAM and decreased autophagy	Qu *et al.* [79]

BC: Breast cancer, EC: Endometrial cancer, CC: Cervical cancer, PTC: Papillary thyroid carcinoma, HCC: Hepatocellular carcinoma, LC: Laryngeal Cancer, GC: Gastric Cancer, NSCLC: Non-small Cell Lung Cancer, CoC: Colorectal Cancer, GBM: Glioblastoma.

Conclusion and Perspectives

Membrane-associated ER-α36 is a small isoform of ER-α that mainly mediates the non-genomic effects of ER regulation in various pathological processes. It plays a different role from traditional ERs. The expression of ER-α36 has been documented in many human cancers, and it appears that no matter whether the cancer is female or male dominated, ER-α36 functions to facilitate the development and/or growth of the cancers in most cases. Therefore, it should be considered as an oncogene.

The clinical significance of ER-α36 needs further exploration. The therapeutic impact of ER-α36 is hindered by the fact that we do not have a specific inhibitor of ER-α36. Furthermore, the detailed signaling pathway of ER-α36 is also not fully known. Further research should focus on the development of new drugs by using the structural and biological differences between ER-α36 and traditional ERs. A new therapy that specifically targets ER-α36 may solve not only the problem of the side effects of existing selective ER regulators but also the resistance that is frequently seen in the current anti-estrogen treatments for certain cancers.

Acknowledgments

This study was supported by grants from the National Natural Science Foundation of China (No.81972493) and the Research Grants Council of the Hong Kong Special Administrative Region (Project No. GRF 14109716).

References

[1] Nilsson S and Gustafsson JA. *Crit Rev Eukaryot Gene Expr.* 2002; 12(4), 237-57.
[2] Cunha GR, Donjacour AA, Cooke PS, Mee S, Bigsby RM, Higgins SJ, Sugimura Y. *Endocr Rev.* 1987; 8(3), 338-62.
[3] Shao W and Brown M. *Breast Cancer Res.* 2004; 6(1), 39-52.
[4] Sun L, Wang J, Zhang L, Li X, Shen D. *Gynecol Obstet Invest.* 2013; 75(1), 68-72.
[5] Sukocheva OA. *World J Gastroenterol.* 2018; 24(1), 1-4.
[6] Sołtysik K and Czekaj P. *Eur J Cell Biol.* 2015; 94(12), 611-25.
[7] Wang ZY and Yin L. *Mol Cell Endocrinol.* 2015; 418 Pt 3, 193-206.
[8] Green S, Walter P, Kumar V, Krust A, Bornert JM, Argos P, Chambon P. *Nature.* 1986; 320(6058), 134-9.
[9] Kuiper GG, Enmark E, Pelto-Huikko M, Nilsson S, Gustafsson JA. *Proc Natl Acad Sci U S A.* 1996; 93(12), 5925-30.
[10] Walter P, Green S, Greene G, Krust A, Bornert JM, Jeltsch JM, Staub A, Jensen E, Scrace G, Waterfield M, *et al. Proc Natl Acad Sci U S A.* 1985; 82(23), 7889-93.

[11] Li L, Haynes MP, Bender JR. *Proc Natl Acad Sci U S A.* 2003; 100(8), 4807-12.

[12] Wang Z, Zhang X, Shen P, Loggie BW, Chang Y, Deuel TF. *Biochem Biophys Res Commun.* 2005; 336(4), 1023-7.

[13] Kang L, Zhang X, Xie Y, Tu Y, Wang D, Liu Z, Wang ZY. *Mol Endocrinol. 2010*; 24(4), 709-21.

[14] Yin L, Zhang XT, Bian XW, Guo YM, Wang ZY. *PLoS One.* 2014; 9(9), e107369.

[15] Flouriot G, Brand H, Denger S, Metivier R, Kos M, Reid G, Sonntag-Buck V, Gannon F. *EMBO J.* 2000; 19(17), 4688-700.

[16] Li H, An LX, Gu YQ. *Progress in Pharmaceutical Sciences.* 2010; 34(2), 61-67.

[17] Rao J, Jiang X, Wang Y, Chen B. *J Steroid Biochem Mol Biol.* 2011; 127(3-5), 231-7.

[18] Yin L and Wang ZY. *Steroids.* 2016; 111, 95-99.

[19] Wang Z, Zhang X, Shen P, Loggie BW, Chang Y, Deuel TF. *Proc Natl Acad Sci U S A.* 2006; 103(24), 9063-8.

[20] Feng S, Wang Y, Wang X, Wang Z, Cui Y, Liu J, Zhao C, Jin M, Zou W. *Sci China Life Sci.* 2010; 53(5), 598-605.

[21] Díaz-Chico BN, Ogasawara Y, Chamness GC, Salman M, McGuire WL. *J Steroid Biochem.* 1988; 30(1-6), 315-20.

[22] Kim KH and Bender JR. *Sci STKE.* 2005; 2005(288), pe28.

[23] Lee LM, Cao J, Deng H, Chen P, Gatalica Z, Wang ZY. *Anticancer Res.* 2008; 28(1B), 479-83.

[24] Chaudhri RA, Schwartz N, Elbaradie K, Schwartz Z, Boyan BD. *Steroids.* 2014; 81, 74-80.

[25] Lin SL, Yan LY, Liang XW, Wang ZB, Wang ZY, Qiao J, Schatten H, Sun QY. *Reprod Biol Endocrinol.* 2009; 7, 102.

[26] Zhang X, Deng H, Wang ZY. *J Steroid Biochem Mol Biol.* 2014; 143, 434-43.

[27] Zhang X, Ding L, Kang L, Wang ZY. *PLoS One.* 2012; 7(1), e30174.

[28] Tu BB, Lin SL, Yan LY, Wang ZY, Sun QY, Qiao J. *Am J Obstet Gynecol.* 2011; 205(3), 227.e1-6.

[29] Zhang XT, Kang LG, Ding L, Vranic S, Gatalica Z, Wang ZY. *Oncogene.* 2011; 30(7), 770-80.

[30] Anastasiadi Z, Lianos GD, Ignatiadou E, Harissis HV, Mitsis M. *Updates Surg.* 2017; 69(3), 313-317.

[31] Weiss A, Noorbakhsh A, Tokin C, Chang D, Blair SL. *Ann Surg Oncol.* 2013; 20(10), 3274-8.

[32] Symmans WF, Hatzis C, Sotiriou C, Andre F, Peintinger F, Regitnig P, Daxenbichler G, Desmedt C, Domont J, Marth C, Delaloge S, Bauernhofer T, Valero V, Booser DJ, Hortobagyi GN, Pusztai L. *J Clin Oncol.* 2010; 28(27), 4111-9.

[33] Lumachi F, Santeufemia DA, Basso SM. *World J Biol Chem.* 2015; 6(3), 231-9.

[34] García-Becerra R, Santos N, Díaz L, Camacho J. *Int J Mol Sci.* 2012; 14(1), 108-45.

[35] Cheng SB, Quinn JA, Graeber CT, Filardo EJ. *J Biol Chem.* 2011; 286(25), 22441-55.

[36] Shi L, Dong B, Li Z, Lu Y, Ouyang T, Li J, Wang T, Fan Z, Fan T, Lin B, Wang Z, Xie Y. *J Clin Oncol.* 2009; 27(21), 3423-9.

[37] Kabos P, Finlay-Schultz J, Li C, Kline E, Finlayson C, Wisell J, Manuel CA, Edgerton SM, Harrell JC, Elias A, Sartorius CA. *Breast Cancer Res Treat.* 2012; 135(2), 415-32.

[38] Lin SL, Yan LY, Zhang XT, Yuan J, Li M, Qiao J, Wang ZY, Sun QY. *PLoS One.* 2010; 5(2), e9013.

[39] Xia B, Tian C, Guo S, Zhang L, Zhao D, Qu F, Zhao W, Wang Y, Wu X, Da W, Wei S, Zhang Y. *Leuk Res.* 2015; 39(1), 92-9.

[40] Si W, Shen J, Zheng H, Fan W. 2019; 11(1), 25.

[41] Simões BM, O'Brien CS, Eyre R, Silva A, Yu L, Sarmiento-Castro A, Alférez DG, Spence K, Santiago-Gómez A, Chemi F, Acar A, Gandhi A, Howell A, Brennan K, Rydén L, Catalano S, Andó S, Gee J, Ucar A, Sims AH, Marangoni E, Farnie G, Landberg G, Howell SJ, Clarke RB. *Cell Rep.* 2015; 12(12), 1968-77.

[42] Deng H, Yin L, Zhang XT, Liu LJ, Wang ML, Wang ZY. *J Steroid Biochem Mol Biol.* 2014; 144 Pt B, 417-26.

[43] Wang Q, Jiang J, Ying G, Xie XQ, Zhang X, Xu W, Zhang X, Song E, Bu H, Ping YF, Yao XH, Wang B, Xu S, Yan ZX, Tai Y, Hu B, Qi X, Wang YX, He ZC, Wang Y, Wang JM, Cui YH, Chen F, Meng K, Wang Z, Bian XW. *Cell Res.* 2018; 28(3), 336-358.

[44] Tong JS, Zhang QH, Wang ZB, Li S, Yang CR, Fu XQ, Hou Y, Wang ZY, Sheng J, Sun QY. *PLoS One.* 2010; 5(11), e15408.

[45] Moreno V, Bosch FX, Muñoz N, Meijer CJ, Shah KV, Walboomers JM, Herrero R, Franceschi S; International Agency for Research on Cancer. Multicentric Cervical Cancer Study Group. *Lancet.* 2002; 359(9312), 1085-92.

[46] Chung SH, Wiedmeyer K, Shai A, Korach KS, Lambert PF. *Cancer Res.* 2008; 68(23), 9928-34.

[47] Sun Q, Liang Y, Zhang T, Wang K, Yang X. *Biochem Biophys Res Commun.* 2017; 487(3), 625-632.

[48] Zhang Wenjin, Peng Yao, Jin ran, *et al. J Tianjin Univ Tradit Chin Med,* 2012; 31 (1), 23-26

[49] Rubio GA, Catanuto P, Glassberg MK, Lew JI, Elliot SJ. *Surgery.* 2018; 163(1), 143-149.

[50] Derwahl M and Nicula D. *Endocr Relat Cancer.* 2014; 21(5), T273-83.

[51] Hima S and Sreeja S. *IUBMB Life.* 2016; 68(2), 85-96.

[52] Przybylik-Mazurek E, Hubalewska-Dydejczyk A, Fedorowicz A, Pach D. *Gynecol Endocrinol.* 2012; 28(2), 150-5.

[53] Holland J, Bandelow S, Hogervorst E. *Maturitas.* 2011; 69(4), 322-37.

[54] Dai YJ, Qiu YB, Jiang R, Xu M, Zhao L, Chen GG, Liu ZM. *Sci Rep.* 2017; 7(1), 12279.

[55] Yang XS, Zhao JW, Zhao SS, Zhou i, Zhong GX, Cheng SM, Zhang IP, and Ding HL. *Chinese Journal of Anatomy.* 2014; 37(5), 614-15.

[56] Bowen RL, Isley JP, Atkinson RL. *J Neuroendocrinol.* 2000; 12(4), 351-4.

[57] Kumar SK, Callander NS, Alsina M, Atanackovic D, Biermann JS, Castillo J, Chandler JC, Costello C, Faiman M, Fung HC, Godby K, Hofmeister C, Holmberg L, Holstein S, Huff CA, Kang Y, Kassim

A, Liedtke M, Malek E, Martin T, Neppalli VT, Omel J, Raje N, Singhal S, Somlo G, Stockerl-Goldstein K, Weber D, Yahalom J, Kumar R, Shead DA. *J Natl Compr Canc Netw.* 2018; 16(1), 11-20.

[58] Miceli V, Cocciadiferro L, Fregapane M, Zarcone M, Montalto G, Polito LM, Agostara B, Granata OM, Carruba G. *OMICS.* 2011; 15(5), 313-7.

[59] Zheng D, Wang X, Antonson P, Gustafsson JÅ, Li Z. *Mol Cell Endocrinol.* 2018; 471, 33-41.

[60] Hassan MM, Botrus G, Abdel-Wahab R, Wolff RA, Li D, Tweardy D, Phan AT, Hawk E, Javle M, Lee JS, Torres HA, Rashid A, Lenzi R, Hassabo HM, Abaza Y, Shalaby AS, Lacin S, Morris J, Patt YZ, Amos CI, Khaderi SA, Goss JA, Jalal PK, Kaseb AO. *Clin Gastroenterol Hepatol. 2017*; 15(11), 1791-1799.

[61] Chen YF, Velmurugan BK, Wang HL, Tu CC, Che RJ, Chen MC, Jen LB, Vishwanadha VP, Hsu HH, Huang CY. *Environ Toxicol.* 2017; 32(2), 519-529.

[62] Zhang J, Ren J, Wei J, Chong CC, Yang D, He Y, Chen GG, Lai PB. *BMC Cancer.* 2016; 16(1), 926.

[63] Cocciadiferro L, Miceli V, Granata OM, Carruba G. *J Steroid Biochem Mol Biol.* 2017; 172, 222-230.

[64] Lau MY, Han H, Hu J, Ji C. *J Gastroenterol Hepatol.* 2013; 28(3), 576-83.

[65] Teng Y, Radde BN, Litchfield LM, Ivanova MM, Prough RA, Clark BJ, Doll MA, Hein DW, Klinge CM. *J Biol Chem.* 2015; 290(25), 15799-811.

[66] Chen J, Chen L, Lu T, Xie Y, Li C, Jia Z, Cao J. *Int J Clin Exp Pathol.* 2019; 12(9), 3222-3234.

[67] Shi L, Feng Y, Lin H, Ma R, Cai X. *J Transl Med.* 2014; 12, 93.

[68] Chai L, Liu H, Zhang Z, Wang F, Wang Q, Zhou S, Wang S. *Tohoku J Exp Med.* 2014; 232(1), 9-19.

[69] Cook MB, Dawsey SM, Freedman ND, Inskip PD, Wichner SM, Quraishi SM, Devesa SS, McGlynn KA. *Cancer Epidemiol Biomarkers Prev.* 2009; 18(4), 1174-82.

[70] Schwartz N, Chaudhri RA, Hadadi A, Schwartz Z, Boyan BD. *Horm Cancer.* 2014; 5(1), 22-32.

[71] Torre LA, Bray F, Siegel RL, Ferlay J, Lortet-Tieulent J, Jemal A. *CA Cancer J Clin.* 2015; 65(2),87-108.

[72] Sipponen P and Correa P. *Gastric Cancer.* 2002; 5(4), 213-9

[73] Camargo MC, Goto Y, Zabaleta J, Morgan DR, Correa P, Rabkin CS. *Cancer Epidemiol Biomarkers Prev.* 2012; 21(1), 20-38.

[74] Lindblad M, Ye W, Rubio C, Lagergren J. *Cancer Epidemiol Biomarkers Prev.* 2004; 13(12), 2203-7.

[75] Deng H, Huang X, Fan J, Wang L, Xia Q, Yang X, Wang Z, Liu L. *Oncol Rep.* 2010; 24(1), 171-6.

[76] Wang J, Li J, Fang R, Xie S, Wang L, Xu C. *Oncol Lett.* 2012; 3(1), 172-175.

[77] Zou F, Wang XM, and Liu LJ. *Chinese Journal of Pathophysiology.* 2012; 28(12), 2167-171.

[78] Fu Z, Wang X, Wang Z, Liu L. *Oncol Lett.* 2018; 15(6), 10031-10036.

[79] Wang X, Huang X, Fu Z, Zou F, Li Y, Wang Z, Liu L. *Int J Oncol.* 2014; 45(6), 2325-30.

[80] Segawa T, Shozu M, Murakami K, Kasai T, Shinohara K, Nomura K, Ohno S, Inoue M. *Clin Cancer Res.* 2005; 11(6), 2188-94.

[81] Han WD, Si YL, Zhao YL, Li Q, Wu ZQ, Hao HJ, Song HJ. *J Steroid Biochem Mol Biol.* 2008 Mar;109(1-2):47-56.

[82] Zhang S, Qiu C, Wang L, Liu Q, Du J. *Steroids.* 2014; 87, 39-45.

[83] Ries LA, Wingo PA, Miller DS, Howe HL, Weir HK, Rosenberg HM, Vernon SW, Cronin K, Edwards BK. *Cancer.* 2000; 88(10), 2398-424.

[84] DeCosse JJ, Ngoi SS, Jacobson JS, Cennerazzo WJ. *Eur J Cancer Prev.* 1993; 2(2), 105-15.

[85] Nanda K, Bastian LA, Hasselblad V, Simel DL. *Obstet Gynecol.* 1999; 93(5 Pt 2, 880-8.

[86] Paganini-Hill A. *Dis Colon Rectum.* 1999; 42(10), 1300-5.

[87] Jiang H, Teng R, Wang Q, Zhang X, Wang H, Wang Z, Cao J, Teng L. *J Steroid Biochem Mol Biol.* 2008; 112(1-3), 20-4.

[88] Spence AM, Peterson RA, Scharnhorst JD, Silbergeld DL, Rostomily RC. *J Neurooncol.* 2004; 70(1), 91-5.

[89] Teng J, Hejazi S, Hiddingh L, Carvalho L, de Gooijer MC, Wakimoto H, Barazas M, Tannous M, Chi AS, Noske DP, Wesseling P, Wurdinger T, Batchelor TT, Tannous BA. *Neuro Oncol.* 2018; 20(5), 642-654.

[90] Pollack IF, DaRosso RC, Robertson PL, Jakacki RL, Mirro JR Jr, Blatt J, Nicholson S, Packer RJ, Allen JC, Cisneros A, Jordan VC. *Clin Cancer Res.* 1997; 3(7), 1109-15.

[91] Puchner MJ and Giese A. *Int J Cancer.* 2000; 86(4), 468-73.

[92] Feng Y, Huang J, Ding Y, Xie F, Shen X. *Oncol Rep.* 2010; 24(6), 1561-7.

[93] Liu Y, Huang L, Guan X, Li H, Zhang QQ, Han C, Wang YJ, Wang C, Zhang Y, Qu C, Liu J, Zou W. *Steroids.* 2016; 111, 127-133.

[94] Qu C, Ma J, Zhang Y, Han C, Huang L, Shen L, Li H, Wang X, Liu J, Zou W. *Cancer Sci.* 2019; 110(1), 221-234.

[95] Zheng Y, Zhang J, Xu ZZ, Sheng JM, Zhang XC, Wang HH, Teng XD, Liu XJ, Cao J, Teng LS. *J Zhejiang Univ Sci B.* 2010; 11(2), 144-50.

[96] Vranic S, Gatalica Z, Deng H, Frkovic-Grazio S, Lee LM, Gurjeva O, Wang ZY. *J Clin Pathol.* 2011; 64(1), 54-7.

In: Estrogen Receptors
Editors: George G Chen et al.
ISBN: 978-1-53618-228-6

Chapter 8

ESTROGEN RECEPTORS AND OBESITY, WITH FOCUS ON ADIPOSE TISSUES

Arvand Asghari[1] and Michihisa Umetani[1,2,*], PhD

[1]Center for Nuclear Receptors and Cell Signaling,
University of Houston, Houston, TX, US
[2]HEALTH Research Institute, University of Houston, Houston, TX, US

ABSTRACT

Obesity, or an increase in general adiposity, is defined as body mass index (BMI) of more than 30 kg/m^2 and it is estimated that more than 35% of US people are obese, while more than half of the people are overweight/obese (BMI $\geq$ 25 kg/m^2). However, due to the complexity and multifactorial nature, the precise mechanisms of obesity are not yet fully understood, and the relations between obesity and diseases other than metabolic diseases, such as cancer, are not yet clear. Estrogen receptors (ERs) in adipose tissues regulate fuel homeostasis and protect from body weight gain, which makes them important targets in obesity and obesity-related diseases. Here, we will discuss the role of ERs in adipose tissue formation and function, and also the recent findings on endogenous and exogenous selective estrogen receptor modulators (SERMs) and their

* Corresponding Author's Email: mumetani@uh.edu.

actions in adipose tissues. We will also discuss how these ER ligands promote or inhibit obesity in different contexts.

Keywords: estrogen, estrogen receptor, adipose, obesity, SERM, 27-hydroxycholesterol, brown adipose tissue, white adipose tissue, energy homeostasis

INTRODUCTION

Adipose tissue, commonly known as "fat" tissue, is a loose connective tissue made up of lipid-filled cells (adipocytes) surrounded by a matrix of collagen fibers, blood vessels, fibroblasts, and immune cells. This tissue can be divided into two very different and functionally opposite depots, namely white adipose tissue (WAT) and brown adipose tissue (BAT). While WAT was known in mammals for centuries, BAT was recognized back in 1551 [1]. Before the 20th century, there was not much interest in adipose tissue as it was considered only as a connective tissue with some lipids inside it, but this point of view was significantly changed in the past century. White adipose tissue is prominently known as an energy storage site mostly due to its limitless capacity in storing triglycerides [2]. However, this tissue is way more than just a reservoir for energy storage. Around 1987, it was found that adipose tissue is a major site for the metabolism of steroids [3], and by 1994, leptin and its function was identified and characterized in adipose tissue [4], which altogether, convinced the scientific community that adipose tissue also works as an endocrine organ [5]. Brown adipose tissue, on the other hand, is mainly responsible for burning the triglycerides and generating heat by uncoupling the mitochondrial fatty acid oxidation. Furthermore, similar to WAT, brown adipose tissue functions as an endocrine organ as well, secreting factors such as fibroblast growth factor 21 [6].

Although the importance of estrogen in adipose tissues has been recognized for a long time, the presence of estrogen receptors in adipose tissues was controversial [7]. By the last decade of the 20th century, the scientific community realized that two ER isoforms, ERα and ERβ, are expressed in the adipose tissue of humans and rodents [7, 8]. Here, we will review the role of estrogen, ER, and their modulators in WAT and BAT, and also explore the potential futures of the estrogen receptor research in adipose tissues.

WHITE ADIPOSE TISSUE

Obesity, or an increase in general adiposity, is defined as body mass index (BMI) of more than 30 kg/m^2, and it is directly related to WAT mass inside the body [9]. In 2016, it was estimated that more than 2 billion people around the world are either overweight or obese [9], while in the United States, more than 39% of adults were estimated as obese [10]. With obesity rates skyrocketing in the United States and globally, the concerns over obesity-related diseases are growing quickly, which stimulates immense research interests on WAT and its functions.

White adipose tissue is a soft tissue that is found in dispersed location throughout the body, mainly under the skin, within the abdominal cavity, and in the musculature [11]. The subcutaneous WAT (SWAT), which resides in the upper (deep and superficial abdominal) and lower body (gluteal–femoral), is located under the skin and is responsible for insulation to prevent heat loss, acting as a barrier against dermal infection, and working as a cushion for protection against external mechanical stress [12]. Visceral WAT (VWAT) is anatomically located in the visceral region (omental, mesenteric, mediastinal and epicardial) buried around vital organs within the peritoneum and rib cage [12]. Moreover, the adipocytes in WAT tend to have a single lipid droplet and therefore are referred to as univacuolar (or unilocular) adipocytes. Interestingly, since the size of the lipid droplet can vary significantly, correlating well with the degree of obesity, the size of the white adipocytes varies considerably as well, ranging from 30 to 150μm [13]. Nevertheless, it is noteworthy that adipose tissue is not only a cluster of adipocytes, but it accommodates a variety of cells, namely, fibroblasts, endothelial cells, and macrophages which account for around 50% of the total cellular content of the adipose tissue [14].

WHITE ADIPOSE TISSUE FUNCTION

The classic role of WAT is to maintain the energy homeostasis, which is to store the energy in form of lipids (triglycerides) and to release them in form of non-esterified fatty acids for oxidation by other organs if necessary. Basically, the size of the adipose tissue and lipid droplets inside adipocytes increases in time of positive energy balance, and decreases in periods that energy expenditure surpasses the available energy [12]. In obese individuals, WAT expands significantly in size, and this expansion comes with various other problems, such as inflammation in the tissue, which we will discuss later on. However, maintaining the energy homeostasis is not the only role of WAT in mammals. As described above, WAT also works as a thermal insulator and a protector of other organs from mechanical damage [14]. Subcutaneous WAT is mainly responsible for the thermal insulation of the body, but the importance of its function varies based on the location. In the trunk, subcutaneous fat and skin are responsible for the majority of the internal insulation, whereas, in the limbs, the underlying musculature plays a very significant role alongside skin and fat in providing insulation, and in hands and feet, WAT plays a minor role in the thermal protection from cold [15]. Yet, the important functions of WAT is not limited to those mentioned above.

As early as 1987, scientists realized that WAT is more than just a repository for lipids. The characterization of adipose tissue as the major site of sex steroids metabolism [3] paired with the identification of adipsin production in adipose tissue, brought the idea that adipose tissue can be a potential endocrine organ [16]. Then, the identification and characterization of leptin in 1994 confirmed the position of WAT as an endocrine organ [4]. Nowadays, WAT is known to secrete a variety of adipokines that can affect the metabolism, immune responses, and signaling pathways throughout the body through endocrine and paracrine/autocrine signaling. Among the various adipokines that WAT secretes, leptin acts on the brain and modulates the appetite, while adiponectin alongside adipsin work on the pancreas and regulate the insulin homeostasis, and leptin and adiponectin work in harmony to affect the cardiovascular system [5]. The integral role of WAT in the body metabolism and hormone signaling has been further emphasized by findings that both deficiency and surplus of WAT is detrimental to the entire body in humans and rodents. Early research on transgenic mice lacking WAT (lipodystrophic animals) showed that they are diabetic and exhibiting both hyperglycemia and substantial hyperinsulinemia [17]. On the other hand, research on obese individuals showed that excess adipose tissue is associated with insulin resistance, hyperglycemia, dyslipidemia, hypertension, and prothrombotic and proinflammatory states [18]. All these findings expanded the role of WAT in the body as a multifunctional tissue responsible for maintaining the energy homeostasis in the body, as well as glucose homeostasis, insulin homeostasis, and several other functions through the endocrine activities of the tissue. This makes WAT one of the most important tissues to study in the body.

Estrogen Actions in White Adipose Tissue

The relationship between estrogen and adipose tissues is reciprocal. Adipose tissue is one of the major sources of estrogen production, and estrogen is one of the biggest modulators of adipose tissue development and function. Back in 1974, Hemsell et al. reported for the first time that adipose tissue is a major source of estrogen production and that the level of estrogen production increases with age in both men and women [19]. The production of estrogens is also increased upon obesity and after menopause, further emphasizing the role of adipose tissue in estrogen production [20].

Exploring the impacts of estrogen on adipose tissue function started around the time that reports suggested a crucial role for adipose tissues in the production of estrogens. However, it is noteworthy to know that most of the early research around this topic was limited to studies on females, as they have significantly higher levels of circulating estrogens produced in the ovary before menopause. The observation that ovariectomy causes obesity and also that the 17β-estradiol (E_2) treatment on ovariectomized rats protects against obesity was a milestone in establishing a remarkable connection between estrogens and adipose tissues [21]. Later on, the similar effect was reported in postmenopausal women, where exogenous estrogen decreased the elevated WAT mass [22]. Moreover, it was reported that estrogens regulate the catalytic activity and mRNA expression of lipoprotein lipase, the rate-limiting enzyme controlling the lipid storage process in both men and women [23, 24].

As early as the 1970s, researchers realized that estrogens also play an important role in glucose homeostasis in females. It was initiated from the observation that glucose tolerance and insulin secretion in female mice is closely related to the estrous cycle [25]. Moreover, ovariectomy in rodents impaired glucose tolerance, caused hyperglycemia, and modulated the glucose-induced insulin secretion, while estrogen treatment could prevent the effects [26]. All these observations in rodents match with the phenotypes in humans, where estrogens were able to reverse the reduced insulin sensitivity in postmenopausal women [22, 27]. These reports emphasized another potential role of estrogens in adipose tissues and modulation of glucose homeostasis. At first, researchers believed that the antiadipogenic actions of estrogens on adipose tissues were mostly through indirect effects, such as decreasing food intake [28] and increasing voluntary exercise [29]. However, further studies showed a direct involvement of estrogens in the cytosol [30] and nucleus [31] of rat adipocytes, suggesting the presence of functional ER in rat adipose tissues.

ESTROGEN RECEPTOR EXPRESSION IN HUMAN WAT

Following the finding of ER in adipose tissues of rodents, the focuses turned toward those in human. Several groups examined the presence of ERα and ERβ in human adipose tissues. Mizutani et al. reported the presence of ER in both protein and mRNA levels in human SWAT [7], while later on Pedersen et al. reported that the level of ER expression seems to be different based on fat depots. Women had the same ER expression in the subcutaneous abdominal region compared to visceral depots, whereas men had twice as high expression in SWAT compared to visceral fat depots [32]. Following the identification and characterization of ERβ, Crandall et al.

further confirmed the presence of this isoform of ER in human SWAT as well [33].

However, by the early 2000s, the expression of ERs in the human adipose tissue remained controversial. While the expression of both estrogen receptors in mature adipocytes had been established, their expression in preadipocytes was under debate. Works by Joyner et al. and Dieudonné et al. revealed the expression of ERα in human preadipocytes and showed that there is no regional or sex differences in its expression, but were not able to detect ERβ in preadipocytes [8,34]. However, other groups detected ERβ expression in preadipocytes, both in mRNA and protein levels, and also reported different levels of expression based on the location and sex [35,36]. Moreover, in addition to mature- and pre-adipocytes, ERα and ERβ are also expressed in other cells found in the adipose tissue, such as vascular endothelial cells, vascular smooth muscle cells, and macrophages. Therefore, we need to consider cells other than adipocytes when talking about the effects of estrogens and ER on adipose tissues [37].

ESTROGEN RECEPTOR FUNCTIONS IN WAT

After the identification of ER in adipose tissues, the next question was about their function in the adipose tissue. One of the most important works in this area was done by Heine et al. in 2000, where they first generated male and female ERα knockout mice (αERKO) and then explored the effects of this deletion in adipose tissues and related phenotypes in mice [38]. In males, while at 5 days postnatal there was no significant increase in the weights of WAT depots in αERKO compared to wild-type (WT) mice, large increases were apparent in ERα knockout males at subsequent ages. At 30 days of age, epididymal fat of αERKO males started to show an increase in weight compared to WT mice, and at 90 days of age, both epididymal and inguinal fat pads were significantly larger in αERKO compared to WT males. The increased WAT in αERKO males compared to WT became more pronounced with advancing age in epididymal, perirenal, and inguinal fat pads. Similar increases in WAT fat pads were observed also in female αERKO mice. Moreover, this increase in the weight of WAT depots was both through hypertrophy, increase in the cell size, and hyperplasia, increase in the cell number. In addition, besides these phenotypic characteristics, there were significant physiological changes in knockout mice compared to WT. Both male and female αERKO mice had insulin resistance and impaired glucose tolerance, similar to humans lacking ERα or aromatase that produces estrogens. Furthermore, while energy intake was equal in WT and αERKO males, energy expenditure was reduced significantly in αERKO compared to WT males [38]. These findings confirmed that estrogen/ERα signaling is critical in female and male WAT, and opened the doors for further studies on ER's roles in adipose tissues.

Around the same time, Ohlsson et al. generated ERβ knockout mice (βERKO) and tried to explore the effects of this deletion in adipose tissues. In their study, βERKO did not alter the body weight, fat distribution, or the serum levels of leptin and insulin. From these results, they concluded that ERβ is not important in adipose tissue metabolism [39]. However, in 2002, Naaz et al. addressed the potential role of ERβ in the adipose tissue using a different approach [40]. The group ovariectomized or sham-operated adult female αERKO mice, fed them with a phytoestrogen-free diet, and treated them with E_2 or vehicle control. While sham-operated αERKO mice had increased body weight, ovariectomized αERKO mice showed a 6% decrease, and E_2 replacement restored body weight to sham levels. Moreover, the weight of fat pads of ovariectomized αERKO mice showed 45% and 16% decrease compared to sham-ovariectomized or E_2-replaced ovariectomized αERKO mice, respectively. Blood glucose levels before and after glucose injection were lower in ovariectomized αERKO mice than sham-operated αERKO mice, and E_2 treatment reversed this phenotype. Insulin levels following glucose challenge were lower in ovariectomized αERKO mice than sham-operated αERKO mice, indicating that ovariectomy ameliorated the glucose intolerance and insulin resistance in αERKO mice. All these observations showed that removing E_2/ERβ signaling in αERKO mice by ovariectomy decreases body and fat-pad weights and adipocyte size, and improves insulin and glucose metabolism. These results showed that ERβ-mediated effects on adipose tissue oppose those of ERα, and E_2 effects on adipose tissue were predominately through ERα.

While it was clear by the early 2000s that ERα signaling pathways are essential for the regulation of energy expenditure and body weight, the underlying mechanisms remained unclear. Originally, ER has been recognized as a transcription factor, and the transcriptional activities are referred to as classical or genomic pathways. In addition, a small amount of ER localized in the cell membrane fraction transmits estrogen-mediated signal transduction independent of its transcriptional activities, and also ER interaction with other transcription factors such as Sp1 and Jun also can lead to indirect transcription of target genes [41, 42]. These are referred to as non-classical or non-genomic functions of ER. During the past couple of years, researchers tried to characterize the non-genomic and noncanonical signaling effects of ERα on adipose tissue by analyzing ERα DNA binding-domain mutant mice. Basically, αERKO mice were used to be introduced with a knock-in of mutant ERα that can only signal through the noncanonical pathway to assess the role of nonclassical ERα signaling in energy homeostasis [43, 44]. These mutant mice showed a restoration of metabolic parameters dysregulated in αERKO mice, such as adiposity, locomotor activity, energy expenditure, and blood glucose levels to near normal values. These results suggest that the nonclassical ERα signaling rescues body weight and metabolic function. Even when mice were put on a high-fat diet, the knock-in ERα DNA binding-domain mutant mice showed similar weight gain to those of WT ($ER\alpha^{+/+}$) mice [44], although on a higher fat diet, the metabolic phenotypes of these mutant female mice were identical with αERKO females [43]. These findings suggest an important and crucial role of ERα in the regulation of metabolic functions in adipose tissues and shed light on the potential role of ER agonists or antagonists as regulators of obesity and obesity-related diseases.

SELECTIVE ESTROGEN RECEPTOR MODULATORS AND ADIPOSE TISSUE

Selective estrogen receptor modulators (SERM) were originally referred to as synthetic non-steroidal agents that act as agonists or antagonists of ER depending on target tissues. To date, only a certain group of oxysterols, metabolites of cholesterol, is identified as endogenous SERM [45]. Among such oxysterols, 27-hydroxycholesterol (27-HC) is the most abundant oxysterol in human circulation, and serum levels of 27-HC correlates well with that of cholesterol [46]. This molecule is generated from cholesterol by a P450 enzyme CYP27A1 and is catabolized by another P450 enzyme CYP7B1 [47]. Interestingly, 27-HC directly binds to both ERα (Ki = 1.32 μM) and ERβ (Ki = 0.42 μM) and exerts its anti-estrogenic effects in the vasculature, while it has pro-estrogenic effects in hepatoma and colon cancer cells [45]. Recently, we investigated the role of 27-HC in adipose tissues and found that 27-HC significantly affects adipose tissue morphology, induces adiposity and weight gain, and also increases the inflammatory responses in adipose tissues regardless of the diet type [48]. To explore the direct effects of 27-HC on adiposity, we fed mice with a high fat, high cholesterol (HFHC) diet or normal chow, and treated them with 27-HC or vehicle control for eight weeks. While 27-HC did not change the body weight gain compared to vehicle treatment with normal chow feeding, mice with 27-HC administrations significantly increased their body weights compared to the control group in the presence of an HFHC diet [48]. Moreover, by using ERα null (αERKO) mice and LXRα/β null (LXRKO) mice, we showed that the increase in adiposity and weight gain upon treatment with 27-HC are dependent on ERα and not on LXR. By performing metabolic studies we confirmed that this weight gain observed by 27-HC administrations was independent of food intake and calorie absorption, and was through the direct effects of 27-HC on WAT. In addition, we found that this increase in WAT weight is due to hyperplasia but not hypertrophy. The induction of hyperplasia by 27-HC was apparent in both SWAT and VWAT depots regardless of the diet type [48]. Interestingly, a high-fat diet alone induces hyperplasia of VWAT only, while estrogen administration induces hyperplasia of both VWAT and SWAT in response to a high-fat diet [49]. In contrast, ovariectomy in female mice decreased hyperplasia in SWAT but not in VWAT in the presence of a high-fat diet. Our results suggest that ER is also involved in the 27-HC-induced hyperplasia in VWAT. Lastly, using gene expression analyses, we found that 27-HC administration up-regulated inflammation-related genes and induced adipose inflammatory responses regardless of the diet, similar to the effects of the HFHC diet alone [48]. The genes induced by 27-HC in a normal chow overlapped with those induced by an HFHC diet, suggesting that 27-HC mimics the inflammatory responses that occur as a consequence of the HFHC diet-feeding. ERα deletion inhibits the beneficial impacts of estrogens on adipose tissues and leads to increased inflammation also in male mice [50]. Therefore, our findings showed that 27-HC inhibits the beneficial effects of estrogens and further increases inflammatory responses via ER in adipose tissues. This is the first study to explore the relationship between 27-HC and adipose tissues, which opens the door toward further exploration of this connection, as inhibition of 27-HC can potentially help to reduce obesity.

As for synthetic SERMs, tamoxifen, a first-generation SERM, blocks ER actions in the breast tissue, while it acts as an ERα agonist in the bone. Thus, tamoxifen is widely used for the treatment of postmenopausal ER-positive breast cancer while simultaneously preventing osteoporosis [51]. However, tamoxifen treatment inhibits the protective effects by estrogens in pancreatic β-cell survival *in vitro* by acting as an ERα antagonist, leading to an increased risk of developing diabetes in breast cancer survivors [52]. Moreover, tamoxifen treatments in mice led to a significant reduction in fat mass but did not change the overall body weight [53], and also led to changing the morphology of WAT and transitioning it toward BAT [54]. Tamoxifen withdrawal leads to fat pad restoration following the acute loss of fat mass, through de novo adipogenesis in VWAT [55]. Raloxifene, another SERM, also acts as an ER agonist in the bone and thus is utilized for the treatment of postmenopausal osteoporosis [56]. In regard to the actions in adipose tissues, raloxifene treatment mimicked the agonistic effects of estrogens on fat mass homeostasis through the modification of leptin and leptin receptor expression [57]. However, clinical studies on postmenopausal women suggested that raloxifene treatment does not follow the effects of estrogens, as it does in mice. While estrogens maintain insulin sensitivity, short- and long-term treatment with raloxifene decreased insulin sensitivity in healthy nondiabetic postmenopausal women and those with postmenopausal osteoporosis, respectively [52].

BROWN ADIPOSE TISSUE

Brown adipose tissue (BAT) is a highly vascularized tissue that contains several cell types including but not limited to brown adipocytes, immune cells, and vascular cells. Brown adipocytes have several small lipid droplets in them (multilocular) and are equipped with multiple mitochondria that give the tissue its brownish color, mainly due to iron contents [58]. Rodents have been known to retain BAT for a long time, but as for humans, it was believed that BAT is only present in newborns and would disappear rapidly in adults. However, in 2007, with the advancements of technology such as positron emission tomography (PET) scanning, scientists started to appreciate the presence of BAT in adult humans as well [59]. The ^{18}F-fluorodeoxyglucose PET locates the tissues of high glucose uptake, and unexpectedly, the regions with high glucose uptake that didn't correspond to any known tissues or tumors were found. Following the analysis of the density and composition of the tissue in those parts using computer tomography, scientists were finally able to come to a solid conclusion that they actually found a new tissue in adult humans, which turned out to be BAT [59].

In rodents and large mammals, classical BAT is primarily located around the interscapular region (iBAT), axillary, paravertebral, and perirenal sites. In humans, it became clear that active BAT is mostly distributed in the upper trunk, such as cervical, supraclavicular, paravertebral, pericardial, and to some extent, mediastinal and mesenteric areas in adults [60]. However, the question of whether BAT is present in all humans or not still remains. Cypess et al. reported that the size of BAT in human varies widely depending on several factors, including but not limited to body mass index, sex, age, environmental temperature, and cholesterol; therefore, it is very hard to detect BAT in adult humans [61].

Brown Adipose Tissue Function

The main and important role of BAT in rodents and newborn humans is to keep the body warm through a process known as non-shivering thermogenesis. This process is initiated via the β3-adrenergic signaling pathway and takes place in numerous mitochondria in brown adipocytes, which contain BAT-specific uncoupling protein, uncoupling protein 1 (UCP1). In order to generate heat, β3-adrenergic receptors in brown adipocytes are activated following the release of norepinephrine from sympathetic nerves, which leads to enhanced transcription of UCP1 and release of fatty acids that enter mitochondria for β-oxidation. UCP1 is a proton channel that uncouples oxidative phosphorylation from ATP production, resulting in the generation of heat instead of ATP. While its exact molecular mechanism is still unclear, UCP1 allows protons to pass from the mitochondrial intermembrane space into the mitochondrial matrix. Activated UCP1, therefore, sets up a futile cycle where the electron transport chain (ETC) pumps protons across the inner mitochondrial membrane, and UCP1 allows them to flow back into the mitochondrial matrix. Thus, the theoretical potential for BAT to oxidize nutrients is limited only by its nutrient supply and capacity for oxidative metabolism; that makes the BAT metabolic rate the highest among any organs [1]. This thermogenic property of BAT is mostly activated in response to cold exposure and is almost not present in thermoneutral conditions; for example, in mice housed at 5°C, 60% of all the energy was expended in BAT [62].

With the identification of BAT in adult humans, several groups started to explore the function of this tissue in adults. Interestingly, studies showed similar trends for BAT activities in response to cold exposure in adult humans, compared to rodents where BAT exhibited significantly higher levels of activity (glucose uptake) in response to cold exposure [63]. Moreover, short-term cold exposure for 10 days in humans not only significantly improved BAT activity, but it also increased whole-body insulin sensitivity by 43% in overweight men, primarily due to increased insulin-stimulated glucose disposal [64]. However, this response varies hugely mediated by several factors, such as sex, age, BMI, and body fat percentage of individuals. Different studies reported that BAT activity is significantly higher in women compared to men, in young people versus the elders, and in lean individuals compared to obese ones [61, 65]. Interestingly, although cold-induced thermogenesis in humans is highly mediated by BAT, there is no evidence to support that diet-induced thermogenesis is also mediated by BAT [66]. Yet, the non-shivering thermogenesis in BAT is well accepted in adult humans, while the degree of importance of BAT in adults is still under debate.

During the past decade, the idea to use the non-shivering thermogenesis of BAT and its capacity to expend energy as a reversal method for cardiac disorders and obesity was emerged [67]. Dong et al. reported that cold-induced BAT activation led to increased plasma levels of small low-density lipoprotein (LDL) remnants and atherosclerotic plaque formation in $ApoE^{-/-}$ and $Ldlr^{-/-}$ mice, which are the model mice of atherosclerosis. In contrast, other groups showed that BAT activation is beneficial for cardiac health and prevents atherosclerosis. [68–70]. This apparent discrepancy comes from the fact that BAT-mediated local lipolysis of triglyceride-rich lipoproteins stimulates the hepatic clearance of lipoprotein remnants via pathways involving the ApoE-LDLR pathway. Since this clearance is not functional in the study by Dong's group, the experimental settings using $ApoE^{-/-}$ and $Ldlr^{-/-}$ mice led to the plaque formation, while in physiological conditions this pathway cleared the remnants and helped to improve cardiovascular health [68]. However, such research is lacking in humans, so whether the same effects can be seen in humans is not yet known. In the context of obesity, as activation of BAT increases the energy expenditure, pharmacological interventions that activate BAT are expected to decrease body weights. However, the results of such pharmacological trials are not good enough yet, because various β3-adrenoreceptor agonists that have been used to date all failed in clinical trials, either due to significant side effects or because of very little impact [71]. Recently, several BAT-derived molecules were found to act in paracrine, autocrine, or endocrine manners. Fibroblast growth factor (FGF) 21, IL-6 and neuregulin 4 are among the first identified BAT-derived factors, so-called batokines, which led to the characterization of BAT not only as a thermogenic organ but also as a secretory organ. Such batokines might direct drug discovery approaches for managing obesity and its associated chronic metabolic diseases [6].

Estrogen, Estrogen Receptors, and Brown Adipose Tissue

Although the effects of estrogens and ER on WAT have been studied extensively, there has not been much data accumulated yet on the impacts of estrogens and ER on BAT. As previously discussed, ovariectomy in rats increases the body weight, primarily through overeating, and the effect can be reversed by estrogen substitution. Early works by several groups suggested that BAT weight, lipolysis, and thermogenesis activities are increased in ovariectomized rats upon treatment with estrogens [72, 73]. Later on, Pedersen et al. showed that UCP1 expression was significantly higher in the BAT of intact rats or the ones with ovariectomy followed by estrogen-substitution compared to the rats with ovariectomy alone, suggesting a potential role of estrogens in BAT morphology and function [74]. However, in cell culture models, estrogens suppressed or did not change the expression of UCP1, while decreasing the proliferation of brown adipocytes in a dose-dependent manner. These results conflict with the findings from animal studies [75, 76]. Further studies showed that estrogens also regulate the expression of adrenergic receptors and mitochondrial biogenesis-signaling factors, such as phosphatase and tensin homolog (PTEN), nuclear respiratory transcriptional factor 1 (NRF1) and possibly peroxisome proliferator-activated receptor g coactivator-1α (PGC-1α) in BAT [77–79]. These findings suggest estrogens regulate BAT in various complicated manners.

Of note, the majority of the effects by estrogens on BAT were associated with increased levels of norepinephrine turnover, which was reduced after ovariectomy. Several groups looked at the modulatory role of estrogens on the brain and sympathetic nerves discharge to BAT [80, 81]. The ventromedial nucleus of the hypothalamus (VMH) is the center for regulating energy expenditure, in particular thermoregulation. Electrical, pharmacological and hormonal stimulation of VMH was reported to increase the temperature of iBAT through the central nervous system [82–84]. In 2007, Musatov et al. first reported a very crucial finding that silencing ERα within the VMH induces obesity, hyperglycemia, and reduced energy expenditure in both mice and rats [85]. Later on, Xu et al. reported that distinct neurons in the hypothalamus regulate estrogen's effects on food intake and energy expenditure and that ERα deletion in the SF1 neurons of VMH led to reduced thermogenic activities of BAT by suppressing UCP1 expression and increased lipid accumulation in gonadal WAT [86]. More recently, researchers were able to demonstrate that E_2 inhibits AMP-activated protein kinase (AMPK) via ERα selectively in the VMH, leading to the activation of thermogenesis in BAT through the sympathetic nervous system in a feeding-independent manner [87].

The expression of ER in BAT is known from the 1970s [88], but the importance of these receptors and their direct interaction with estrogens in BAT is still unclear. The deletion of ERα did not lead to any effect on BAT weight [38]. However, administration of an ERβ-selective agonist to high-fat diet-fed female mice increased the expression of thermogenic UCP1 in BAT, thereby reducing obesity, although the possibility that the compound acted through other mechanisms than ERβ modulation remains [89]. While the genomic function of estrogens is mediated by both ERα and ERβ, some non-genomic signaling pathways by estrogens seem to be mediated by G protein-coupled estrogen receptor (GPER) [90]. GPER is found on the plasma membrane and membrane of endoplasmic reticulum, and is generally responsible for the production of cyclic adenosine monophosphate, transactivation of epidermal growth factor (EGFR), and activation of multiple kinase pathways [90, 91]. Interestingly, both male and female mice lacking GPER developed moderate obesity with significant reductions in energy expenditure, increased BAT lipid content, and decreased expression of UCP1 and β3-adrenergic receptor, indicating reduced thermogenic activities [92]. Similar to rodents, both ERα and ERβ are expressed with ERα being more abundant in human BAT; yet the importance and effects of these receptors in BAT have not been deeply investigated and remain largely unknown [93].

CONCLUSION

Estrogens and estrogen receptors are of crucial importance for metabolism. Here, we discussed the importance of this hormone and its receptors in obesity and adipose tissue function. Ovariectomy or menopause leads to significant obesity and disruption in glucose homeostasis and energy expenditure [51, 52]. In turn, estrogen administration reverses the effects of ovariectomy or menopause largely by decreasing feeding behavior and increasing energy expenditure [51, 94]. Interestingly, the effects of estrogens are not only by impacting the central nervous system but also through direct interaction with adipose tissues [80]. While the roles of estrogens and ER have been studied extensively in WAT, there is a huge gap in our knowledge regarding the role and importance of these molecules in BAT. Especially for ER, there has been almost no comprehensive study around their role and function in human BAT and how they are connected to important effects of estrogens on energy expenditure and homeostasis. All these unsolved questions demand extensive future studies on the role of estrogens and ER in adipose tissues, as they can be one of the best hopes for addressing the problem of obesity and potential therapeutic approaches toward metabolic diseases.

REFERENCES

[1] Cannon, B., Nedergaard, J., *Physiol. Rev.* 2004; 84, 277.
[2] Ahima, R. S., Flier, J. S., *Trends Endocrinol. Metab.* 2000; 11, 327.
[3] Siiteri, P. K., *Am. J. Clin. Nutr.* 1987; 45, 277.
[4] Zhang, Y., Proenca, R., et al., *Nature* 1994; 372, 425.

[5] Kershaw, E. E., Flier, J. S., *J. Clin. Endocrinol. Metab.* 2004; 89, 2548.

[6] Villarroya, F., Cereijo, R., et al., *Nat. Rev. Endocrinol.* 2017; 13, 26.

[7] Mizutani, T., Nishikawa, Y., et al., *J. Clin. Endocrinol. Metab.* 1994; 78, 950.

[8] Dieudonné, M. N., Leneveu, M. C., et al., *Am. J. Physiol.-Cell Physiol.* 2004; 286, C655.

[9] Smith, K. B., Smith, M. S., *Prim. Care Clin. Off. Pract.* 2016; 43, 121.

[10] Hales, C. M., Carroll, M. D., et al., 2017.

[11] Caballero, B., Ed., *Encyclopedia of food sciences and nutritio*n, Academic Press, Amsterdam, 2003.

[12] Kwok, K. H. M., Lam, K. S. L., et al., *Exp. Mol. Med.* 2016; 48, e215.

[13] Laforest, S., Labrecque, J., et al., *Crit. Rev. Clin. Lab. Sci.* 2015; 52, 301.

[14] Trayhurn, P., *Obes. Rev.* 2007; 8, 41.

[15] Gregory, E. L., *Clin. Dermatol.* 1989; 7, 78.

[16] Flier, J., Cook, K., et al., *Science* 1987; 237, 405.

[17] Moitra, J., Mason, M. M., et al., *Genes Dev.* 1998; 12, 3168.

[18] Grundy, S. M., Brewer, H. B., et al., *Circulation* 2004; 109, 433.

[19] Hemsell, D. L., Grodin, J. M., et al., *J. Clin. Endocrinol. Metab.* 1974; 38, 476.

[20] Nelson, L. R., Bulun, S. E., *J. Am. Acad. Dermatol.* 2001; 45, S116.

[21] Wade, G. N., In *Advances in the Study of Behavior*, Elsevier, 1976, Vol. 6, pp. 201–279.

[22] Tchernof, A., Calles-Escandon, J., et al., *Coron. Artery Dis.* 1998; 9, 503.

[23] Rebuffé-Scrive, M., Lönnroth, P., et al., *Int. J. Obes.* 1987; 11, 347.

[24] Rebuffé-Scrive, M., Andersson, B., et al., *Metabolism* 1989; 38, 453.

[25] Bailey, C. J., Matty, A. J., *Horm. Metab. Res.* 1972; 4, 266.

[26] Bailey, C. J., Ahmed-Sorour, H., *Diabetologia* 1980; 19, 475.

[27] Lindheim, S. R., Buchanan, T. A., et al., *J. Soc. Gynecol. Investig.* 1994; 1, 150.

[28] Tarttelin, M. F., Gorski, R. A., *Acta Endocrinol.* (Copenh.) 1973; 72, 551.

[29] Wade, G. N., *Physiol. Behav.* 1972; 8, 523.

[30] Wade, G. N., Gray, J. M., *Endocrinology* 1978; 103, 1695.
[31] Pedersen, S. B., Børglum, J. D., et al., *Biochim. Biophys. Acta BBA - Mol. Cell Res.* 1991; 1093, 80.
[32] Pedersen, S. B., Hansen, P. S., et al., *Eur. J. Clin. Invest.* 1996; 26, 262.
[33] Crandall, D. L., Busler, D. E., et al., *Biochem. Biophys. Res. Commun.* 1998; 248, 523.
[34] Joyner, J. M., Hutley, L. J., et al., *Endocrine* 2001; 15, 225.
[35] Anwar, A., McTernan, P. G., et al., *Diabetes Obes. Metab.* 2001; 3, 338.
[36] Pedersen, S. B., Bruun, J. M., et al., *Mol. Cell. Endocrinol.* 2001; 182, 27.
[37] Cooke, P. S., Naaz, A., *Exp. Biol. Med.* 2004; 229, 1127.
[38] Heine, P. A., Taylor, J. A., et al., *Proc. Natl. Acad. Sci.* 2000; 97, 12729.
[39] Ohlsson, C., Hellberg, N., et al., *Biochem. Biophys. Res. Commun.* 2000; 278, 640.
[40] Naaz, A., Zakroczymski, M., et al., *Horm. Metab. Res.* 2002; 34, 758.
[41] Acconcia, F., Ascenzi, P., et al., *Mol. Biol. Cell* 2005; 16, 231.
[42] Jakacka, M., Ito, M., et al., *Mol. Endocrinol.* 2002; 16, 2188.
[43] Hart-Unger, S., Arao, Y., et al., *Int. J. Obes.* 2017; 41, 945.
[44] Park, C. J., Zhao, Z., et al., *J. Clin. Invest.* 2011; 121, 604.
[45] Umetani, M., Domoto, H., et al., *Nat. Med.* 2007; 13, 1185.
[46] Brown, A. J., Jessup, W., *Atherosclerosis* 1999; 142, 1.
[47] He, S., Nelson, E. R., *Maturitas* 2017; 104, 29.
[48] Asghari, A., Ishikawa, T., et al., *Endocrinology* 2019; 160, 2485.
[49] Jeffery, E., Wing, A., et al., *Cell Metab.* 2016; 24, 142.
[50] Davis, K. E., D. Neinast, M., et al., *Mol. Metab.* 2013; 2, 227.
[51] Mauvais-Jarvis, F., Clegg, D. J., et al., *Endocr. Rev.* 2013; 34, 309.
[52] Kim, J. H., Cho, H. T., et al., *Endocr. J.* 2014; 61, 1055.
[53] Liu, L., Zou, P., et al., *Cell Death Dis.* 2015; 6, e1586.
[54] Zhao, L., Wang, B., et al., *Int. J. Obes.* 2020; 44, 226.
[55] Ye, R., Wang, Q. A., et al., *Mol. Metab.* 2015; 4, 771.
[56] Ettinger, B., *JAMA* 1999; 282, 637.

[57] Meli, R., Pacilio, M., et al., *Endocrinology* 2004; 145, 3115.
[58] Townsend, K., Tseng, Y.-H., *Adipocyte* 2012; 1, 13.
[59] Nedergaard, J., Bengtsson, T., et al., *Am. J. Physiol.-Endocrinol. Metab.* 2007; 293, E444.
[60] Park, A., *World J. Stem Cells* 2014; 6, 33.
[61] Cypess, A. M., Lehman, S., et al., *N. Engl. J. Med.* 2009; 360, 1509.
[62] Golozoubova, V., Gullberg, H., et al., *Mol. Endocrinol.* 2004; 18, 384.
[63] Orava, J., Nuutila, P., et al., *Cell Metab.* 2011; 14, 272.
[64] Hanssen, M. J. W., Hoeks, J., et al., *Nat. Med.* 2015; 21, 863.
[65] Marken Lichtenbelt, W. D. van, Vanhommerig, J. W., et al., *N. Engl. J. Med.* 2009; 360, 1500.
[66] Peterson, C. M., Lecoultre, V., et al., *Obesity* 2016; 24, 96.
[67] Loh, R. K. C., Kingwell, B. A., et al., *Obes. Rev.* 2017; 18, 1227.
[68] Berbée, J. F. P., Boon, M. R., et al., *Nat. Commun.* 2015; 6, 6356.
[69] Dong, M., Yang, X., et al., *Cell Metab.* 2013; 18, 118.
[70] Laurila, P.-P., Soronen, J., et al., *Sci. Transl. Med.* 2016; 8, 323ra13.
[71] Marlatt, K. L., Ravussin, E., *Curr. Obes. Rep.* 2017; 6, 389.
[72] Kemnitz, J., Glick, Z., et al., *Pharmacol. Biochem. Behav.* 1983; 18, 563.
[73] Richard, D., *Am. J. Physiol.-Regul. Integr. Comp. Physiol.* 1986; 250, R245.
[74] Pedersen, S. B., Bruun, J. M., et al., *Biochem. Biophys. Res. Commun.* 2001; 288, 191.
[75] Quevedo, S., Roca, P., et al., *Pflugers Arch.* 1998; 436, 689.
[76] Picó, C., Puigserver, P., et al., *Mol. Cell. Biochem.* 1998; 189, 1.
[77] Monjo, M., Rodríguez, A. M., et al., *Endocrinology* 2003; 144, 4923.
[78] Rodriguez, A. M., Monjo, M., et al., *Cell. Mol. Life Sci.* 2002; 59, 1714.
[79] Rodríguez-Cuenca, S., Monjo, M., et al., *Am. J. Physiol.-Endocrinol. Metab.* 2007; 292, E340.
[80] López, M., Tena-Sempere, M., *Best Pract. Res. Clin. Endocrinol. Metab.* 2016; 30, 527.
[81] Yoshida, T., Nishioka, H., et al., *Metabolism* 1987; 36, 1.
[82] López, M., Varela, L., et al., *Nat. Med.* 2010; 16, 1001.

[83] Martinez de Morentin, P. B., Whittle, A. J., et al., *Diabetes* 2012; 61, 807.

[84] Perkins, M. N., Rothwell, N. J., et al., *Nature* 1981; 289, 401.

[85] Musatov, S., Chen, W., et al., *Proc. Natl. Acad. Sci.* 2007; 104, 2501.

[86] Xu, Y., Nedungadi, T. P., et al., *Cell Metab.* 2011; 14, 453.

[87] Martínez de Morentin, P. B., González-García, I., et al., *Cell Metab.* 2014; 20, 41.

[88] Wade, G. N., Gray, J. M., *Physiol. Behav.* 1979; 22, 583.

[89] Yepuru, M., Eswaraka, J., et al., *J. Biol. Chem.* 2010; 285, 31292.

[90] Revankar, C. M., *Science* 2005; 307, 1625.

[91] Prossnitz, E. R., Arterburn, J. B., *Pharmacol. Rev.* 2015; 67, 505.

[92] Davis, K. E., Carstens, E. J., et al., *Horm. Behav.* 2014; 66, 196.

[93] Velickovic, K., Cvoro, A., et al., *J. Clin. Endocrinol. Metab.* 2014; 99, 151.

[94] Gao, Q., Horvath, T. L., *Am. J. Physiol.-Endocrinol. Metab.* 2008; 294, E817.

About the Editors

George G Chen
Senior Research Fellow
Department of Otorhinolaryngology, Head and Neck Surgery,
the Chinese University of Hong Kong, Hong Kong
Email: gchen@cuhk.edu.hk

George G Chen is Senior Research Fellow, Department of Otorhinolaryngology, Head and Neck Surgery, Faculty of Medicine, the Chinese University of Hong Kong. He has extensive experience in cancer research, particularly estrogens, estrogen receptors and transcription factors in cancers. He has authored or co-authored 230 papers and has written a number of books or book chapters.

Michael CF Tong
Professor, Chairman
Department of Otorhinolaryngology, Head and Neck Surgery,
the Chinese University of Hong Kong, Hong Kong

Michael CF Tong is Professor and Chairman in Department of Otorhinolaryngology, Head and Neck Surgery, Faculty of Medicine, the

Chinese University of Hong Kong. He is a well-known expert in otorhinolaryngology and actively involving in research in head and neck tumors in the past 20 years in various areas of basic sciences to clinical research. His current role include research planning, administration and supervision of research students.

CA van Hasselt
Professor
Department of Otorhinolaryngology, Head and Neck Surgery, the Chinese University of Hong Kong, Hong Kong

CA van Hasselt is Research Professor in the Department of Otorhinolaryngology, Head and Neck Surgery, Faculty of Medicine, the Chinese University of Hong Kong. He has 35 years of clinical and research experience in otorhinolaryngology, head and neck surgery. He has been a driving force in advancing laboratory research in head and neck cancers and is committed to ensure the research undertaken in the laboratory is related to practical applications that make a difference to patients in clinical practice.

INDEX

C

D

E

F

G

H

I

K

L

M

N

O

P

R

S

T